The Tangled Bank

WRITINGS FROM *Orion*

BOOKS BY ROBERT MICHAEL PYLE

Wintergreen: Rambles in a Ravaged Land
The Thunder Tree : Lessons from an Urban Wildland
Where Bigfoot Walks: Crossing the Dark Divide
Nabokov's Butterflies (with Brian Boyd and Dmitri Nabokov)
Chasing Monarchs: Migrating with the Butterflies of Passage
Walking the High Ridge: Life as Field Trip
Sky Time in Gray's River: Living for Keeps in a Forgotten Place
Mariposa Road: The First Butterfly Big Year
Letting the Flies Out (chapbook: poems, essays, stories)
The Tangled Bank: Writings from Orion

ON ENTOMOLOGY

Watching Washington Butterflies
The Audubon Society Field Guide to North American Butterflies
The IUCN Invertebrate Red Data Book (with S. M. Wells and N. M. Collins)
Handbook for Butterfly Watchers
Butterflies: A Peterson Color-In Book (with Roger Tory Peterson and
   Sarah Anne Hughes)
Insects: A Peterson Field Guide Coloring Book (with Kristin Kest)
The Butterflies of Cascadia

The Tangled Bank

WRITINGS FROM *Orion*

Robert Michael Pyle

Oregon State University Press
Corvallis

All of these essays except "The Fern Wall," "Feathered Remnants," and "X the Unknown" originally appeared in *Orion Afield* or *Orion Magazine*, and are reprinted here with permission.

The paper in this book meets the guidelines for permanence and durability of the Committee on Production Guidelines for Book Longevity of the Council on Library Resources and the minimum requirements of the American National Standard for Permanence of Paper for Printed Library Materials Z39.48-1984.

Library of Congress Cataloging-in-Publication Data
Pyle, Robert Michael.
  The tangled bank : essays from Orion / Robert Michael Pyle.
      p. cm.
  ISBN 978-0-87071-679-9 (alk. paper) -- ISBN 978-0-87071-680-5 (e-book)
  1. Natural history. I. Orion (New York, N.Y.) II. Title.
  QH81.P95 2012
  508--dc23
                                        2012015115

First published in 2012 by Oregon State University Press
Printed in the United States of America

Oregon State University Press
121 The Valley Library
Corvallis OR 97331-4501
541-737-3166 • fax 541-737-3170
http://osupress.oregonstate.edu

For Thea
and
For Jen and Aina: it's your book too

The best piety is to enjoy—when you can. You are doing the most then to save the earth's character as an agreeable planet. And—enjoyment radiates. It is of no use to try and take care of all the world; that is being taken care of when you feel delight—in art or anything else.

—George Eliot, *Middlemarch*

Table of Contents

Prologue: The Fern Wall

For Christmas, Thea gave me a children's book titled *The Riverbank*, beguilingly illustrated by Fabian Negrin, with words by Charles Darwin. Mr. Negrin took as his text the same paragraph that gave the title for this book and for my long-running column in *Orion Afield* and *Orion* magazines upon which it is based. That paragraph is the final one in the *Origin of Species*, in which the author recapitulated his life's work and the entire book's message in 219 words. It begins, "It is interesting to contemplate a tangled bank, clothed with many plants of many kinds, with birds singing on the bushes, with various insects flitting about, and with worms crawling through the damp earth, and to reflect that these elaborately constructed forms, so different from each other, and dependent upon each other in so complex a manner, have all been produced by laws acting around us."

Just after Yule I sat with my grandson Francis and read *The Riverbank* with him, finishing up as does that pithy paragraph: "from so simple a beginning endless forms most beautiful and most wonderful have been, and are being evolved." Francis, four, did not fully understand the words, but he got the message: "So bugs turn into people?" he asked.

For Darwin, the tangled bank he had in mind was presumably a well-vegetated earthen slope above a narrow lane in Shropshire or Kent, such as he had dwelt among for most of his life. But no doubt each of us gains a different mental picture from the phrase. The author and artist of the book I read with Francis clearly pictured Darwin's place of contemplation as the bank of a river. When I first read *The Origin*, not at four but at fourteen, I saw the author's tangle in terms of the banks of my beloved High Line Canal, my own special haunt, overhung with long grasses through which the wood nymphs flitted. Now, as a long-time Northwesterner, nothing reflects the tangled bank in my mind better than a fern wall.

I'd fallen in love with ferns as a boy, both from the stories my grandmother, great-aunt, and mother told from their earlier lives in Washington, and from the few actual ferns I saw in moister mountain enclaves. One particularly rainy summer, when the canyon flooded out on US 6, my mother and brother and I were obliged to overnight in a motel. Out behind the lodge, among the rain-soaked aspens and dripping granite, I came face to face with my first mossy, ferny cliff, and right there in Silver Plume, Colorado, I was hooked. When, that fall, I went off to college at the coast, I found myself surrounded by sword ferns on campus and in every one of the green ravines that cleaved Seattle's hills. On my first trip to the Redwood Country, we hiked into Fern Canyon at Prairie Creek Redwoods State Park. The overwhelming feeling of being surrounded, immersed, all but absorbed by those high lush walls of sheer green growth was something I've known only in a few places, such as Oneonta Gorge off the Columbia River and in Mt. Field National Park, Tasmania. So when I came to live near such a place, here in southwest Washington, I was entirely delighted.

My Fern Wall defines the outside curve of state highway 4 as it swings into Skamokawa. The moss-and-lichen-lined limbs of alders and maples hang a ceiling above it, batted by fronds of cedar and hemlock. Sitka spruces reach out over Skamokawa Creek from the other side. The tidal stream runs beneath the graceful arch of a footbridge before hanging a right to its confluence with the Columbia, hard beneath the luminous wooden tower of Redmen Hall. Raven oversees the scene, and common mergansers drift down the middle. But the main event is the ferns. Masses of robust, kelly green sword ferns spill off the thick-soiled, gentler edges of the slope. Where the wall grows steeper, the lush outburst of the smaller, spring-green licorice ferns takes over. And from its bare-wall vertical, saturated except in summer by dripping seeps, springs the delicate tracery of maidenhair ferns on their black stems above the brown blanket of last year's leavings. Deer and lady ferns grow atop the wall, beneath the Douglas-firs above.

I love driving by the Fern Wall at any season. But as March comes up to April, it really comes into its own. The lushness is such that you want to gnaw and crunch it like salad greens. In June, the goatsbeard will toss out its long creamy danglers. Then summer offers a green shade beneath the umbrella of the hardwoods, as the ferns detonate their sori. When things finally go dry in August, the licorice wilts, goes crispy and curled, and the

Fern Wall turns almost brown. But the autumn rains are never far behind. And with the rain, the *Polypody* pop out again like magic mermaids in a fish bowl. The frost and snow barely touch them. They unfold more and more until, with spring again, the green report becomes an out-and-out outrage.

And that, I think, is what I love most about fern walls: their reliable imperative to fade, then to regroup, to explode, to shock, to clothe the world. But who ever looks, besides me? As log trucks whoosh eastward toward the mills and docks, and cars and pickups head west toward the beach (especially if there's a clam tide), the Fern Wall is just a shady spot to watch for ice on the road, a curve between cliff and river to navigate and be gone, revving up from 35 to the old 55 or 60. Meanwhile, coltsfoot breaks ground and salmonberry's first cerise blooms break out, as winter wren sings under the wall and alumroot keeps its secrets below the soaked *Selaginella*. But maybe I'm wrong, and the Fern Wall gives pleasure and comfort to other drivers too, and to drift boaters among the mergansers on the creek. Maybe it isn't all mine, after all. I *hope* it's not.

The Tangled Bank stands in my mind for the imperishable fascination to be found in the living, physical world. Nothing else, other than love, has ever seemed as necessary to me as intimate connection with that world. And I'm not even sure there's a difference. In *Sky Time in Gray's River*, after listing some of the elements I cherish on my home ground—swallows departing and coming back, the first echo azures of the spring—I wrote that "these things are as important to me as love, and in fact, that's what they are."

A few days after that Christmas visit with Francis, I sat in a coffee house in Astoria, Oregon. Little yellow lamps pooled light on black tables in the December morning's gloom. In one yellow pool, a girl wearing shiny black boots, gray leggings, a black overcoat, and a pink cashmere hood cupped a latte and read the paper as she waited for her mother to bring the pastries. In another cone of lamplight, a woman twinned to her laptop tucked into home fries she'd let get cold. Outside, an old seaman or barman or both glided past the window in a black down parka ten times his bulk, and waterdrops depended from still-sealed maple buds, as steely clouds and rivermist unraveled and a winter-blue sky leaked out. Then full sun came in, swamping the yellow lamps. The whiny oinks of sea lions echoed up from the Columbia River a block away, pulling me out like the songs of sirens. The white cheek-spot of a goldeneye bobbed by the pilings, and the hoods of the buffleheads puffed out as soft as that girl's cashmere.

Every bit as much as a fern wall, a Kentish hedge, or a Borneon rainforest, these are things that I'd call real. Cartesian real. All around, always available, and endlessly engaging to one who is easily amused. But not everyone finds the minutiae of life as interesting as I do. The editorial director of Globe Communications supermarket tabloids, when asked why he published such outlandish fabrications, told the *Los Angeles Times*, "Our readers want to believe this stuff. The world is very boring."

When Charles Darwin posed his "tangled bank, clothed with many plants of many kinds" as a canvas for his brief synopsis of his insights to date, he conjured from it all the elements of the reality that so beguiled him. Out of that hedge, he re-made his sense of the world. And he *changed* the world while he was at it. I have no such aspirations; I merely write to refute the idea that the world is a boring place. I have taken as my subject whatever most wanted to be plucked from the tangle for examination on a given day. The bank, after all, is really the whole show; the tangles are just the most interesting bits.

There's barely a place, scarcely a square inch with any visible life or color to it, that can't catch and keep my interest, at least for a spell. Surely *spell* is the right word, for isn't fascination a species of bewitchment? Recently I stood rapt for a quarter of an hour in a supermarket parking lot, parsing the astonishing dendritic patterns and sensational shimmers of an oily patch of asphalt. Even concrete has its leaf fossils, footprints, and geologic traces. Manhattan's travertine-and-granite canyons throb as much as western ones, seen in the right light. In "The Judgment of the Birds," Loren Eiseley described rock doves viewed from high on such a canyon's rim: "The white city pigeons were beginning to float outward upon the city ... perhaps I had only dreamed about man in this city of wings—which he could surely have never built."

The world is not, after all, boring. The most reliable antidote for ennui is loving attention to the endlessly engaging details of the living earth. I believe it can relieve (if not fix) anyone's life, whether gripped in poverty and suffering or safe and warm behind the walls of privilege. That's easy to say from my lucky vantage; I can't really know anyone's situation but my own. But I do know that Darwin, who knew something of both privilege and despair, found such succor among the earthworms along his beloved Sand Walk. And when he peered into his tangled bank of mystery and revelation, he also found a way—*the* way—to tie all the details together.

Some might ask, where's the *spirit* in all this talk of the material world? But I have never been any more capable of convincingly separating earth from spirit, or spirit from flesh or bone or stone, than humans from the rest of nature; or than I've been able to remain bored for long. To put it another way, nature has never seemed *wanting* to me such as to require a supernature. To me, the world itself is rich and sufficient. It can never be too much with us. And as George Eliot told us, the best way to care for the world, "to save the earth's character as an agreeable planet," is to take delight in it. "The Tangled Bank" was my lucky invitation to share with readers the delight I take in the world every day I live in it.

I was given this chance by *Orion* not once, but twice. The second time took, and you see the results before you. It might not have been thus. Back in the late 1970s, what we know and love now as *Orion* was preceded by a newsprint tabloid known as *Orion Nature Book Review*, edited by Aina Niemela. I was a regular reviewer for it. One day, as I was preparing an omnibus review of several new butterfly books, Aina wrote and told me to keep the books, but not to bother completing the assignment. The publication had gained private backing to become *Orion Nature Quarterly* (later, simply *Orion*), a slick and serious vehicle for examining people and nature. "And by the way," Aina inquired of her reviewers, "let me know what you think this new serial should contain."

I was a fan of Stephen Jay Gould and his column in *Natural History,* "This View of Life" (another phrase, incidentally, drawn from that same paragraph of Darwin's, where he writes "There is grandeur in this view of life"). No doubt with Gould's column in mind, I wrote back to Aina that I felt the new magazine should definitely include a regular voice. I nominated myself to write a column called "The Niche of a Naturalist." To my surprise, Aina and her comrades took me up on it. So it was that "The Niche of a Naturalist" appeared in the first three issues of the new *Orion*. However, the column failed. There were too many cooks around that early pot, each with a different idea of what my column should be: personal, impersonal, more nature, more people, reportage, opinion, and so on. All of them got their editorial licks in, and in the end none of us were happy with the outcome, least of all me. It certainly didn't resemble the engagement with wonder I'd had in mind. So we pulled the plug, and I remained merely an interested observer and reader of *Orion* for the next ten years.

Then in 1992, the dean of American nature writers, John Hay, asked for me to be on the invitation list when he was honored by the Orion folks on Cape Cod. This occasion marked the launch of both the John Hay Medal and the Orion Society. It also furnished my first meeting with Aina, as well as with Marion and Olivia Gilliam, the magazine's publishers. We found there were no hard feelings over the dead-end column, and I became involved again, writing for the magazine and traveling with many of the society's popular Forgotten Language Tours—barnstorming chautauquas of nature writers, *sensu lato*, traveling around a given region giving readings, workshops, and field trips with local writers and lovers of land and literature.

As a member of the Orion Society's Advisory Council, I was invited to attend the annual events around the awarding of the John Hay Medal. In the spring of 1997 we were in Nevada City, California, for presentation of the medal to poet and essayist Gary Snyder. Following the opening gala, Marion Gilliam asked me to take a walk in the balmy Sierra evening. A new young editor, Jennifer Sahn, fresh from Middlebury College where she had studied with John Elder, came with us. Usually when Marion asked to speak with me in private it was to gently upbraid me for going too long at a reading, or some such. But this time was different. He confided in me that the organization was going to launch a companion magazine, to be edited by Jennifer, which would concentrate on community. He asked my opinion of several names under consideration, and I voted strongly for *Orion Afield*. And then he greatly surprised me by saying that a column was wanted to tie the two magazines together through nature, and would I consider writing it? I felt as if I had been rehabilitated after the Cultural Revolution.

So that is how I came to have the remarkable opportunity to indulge my whims, passions, and obsessions in this splendid forum. Jennifer was my editor for every essay except two, when she was on maternity leave. For those I got to work again with Aina Niemela, and the whole fine symmetry of the thing came full circle. This was no dashed-off column. Each essay went through multiple drafts, often five or six; and when it came out, both cooks were happy. The column ran in every issue of *Orion Afield*'s five-year life, at 750 words—a very demanding form, and a whole new education in concision! When *Afield* was folded back into the mother ship in 2003, the column went with it, appearing in *Orion* for another five years at around 1,400 words (only a little less challenging for an author used to 3,000- to 5,000-word essays and book chapters).

"The Tangled Bank" enjoyed an uninterrupted run of fifty-two issues, finishing only because I could not continue writing it during the year I spent *actually* afield on my butterfly big year, for the book *Mariposa Road*. The essays appear here essentially as they were published in the magazines, with only small changes where an infelicity crept in, a correction was needed, or a word or phrase, originally deleted for reasons of space, was restored. One essay that was pulled for internal reasons at *Orion* has been revisited and published at last, in place of another that I have elected to omit on account of substantial shortcomings of my making. Brief notes have been inserted below a few entries in order to bring things up to date or otherwise aid the reader. And so "The Bank" is back. I hope it may touch new readers, and bring a welcome echo to the many devotees who have kindly told me that it has been missed.

Later this spring, with licorice ferns at the peak of eruption and the goatsbeards bursting, I want to take Francis to the Fern Wall. We will stand on the footbridge over Skamokawa Creek, look across at the cliff face, and see how many different kinds of living things we can count, as the log trucks whoosh by. "Yes, Francis," I'll say, "people do come from bugs."

Leaves That Speak

I happen to live in a paradise of leaves. On the whole, the Pacific Northwest, and the maritime rainforest in particular, photosynthesize more with cedar scales and the needles of firs, hemlocks, and spruces than with full-blown, deciduous leaves. But this particular place is an old Swedish farmstead founded by an immigrant, by way of the Midwest, who cared more for horticulture than agriculture. Eventually he returned to Sweden, leaving the farm to dairy-herding descendants for the next seventy years until I came along. But before he sailed home, I I. P. Ahlberg planted and nurtured a remarkable array of European, Midwestern, and native trees and shrubs, several of which are now the largest of their species in the state. The by-product of this fine arboreal legacy is a floristic melting pot of trees from many places reproducing together, a bastard ecosystem that coincidentally spawns, each spring, this paradise of leaves.

As I write, the furrowed broad blades of European hornbeams press toward my study window, overhung by the downy pink unfurlings of the greatest red oak's leaves. The forest beyond grows daily more clogged with the many-greened vanes of English oaks, Swedish birches, and adventitious, exotic sycamore-maples, all growing in company with each native conifer that might be expected on such a site. Along the margins, the freshest spring greens of all express jointly in the luscious tissues of Eurasian beeches and native vine maples and the new growth of Sitka spruces. Then of course there is the panoply of form and verdure of the field layer, the understory, the ground plants, and the chaotic accumulations of more than a century of gardens.

When the people who were here before the Europeans first confronted bibles, books, and treaties, they could see that these strange new objects held great power for their colonial owners. Some of them referred to the black-marked pages as "talking leaves," since they seemed to mediate speech as they turned in the wind like the leaves of trees. Sequoyah, a young

Cherokee man, determined to provide his people with this power, which seemed to come from "making words fast on paper," as he put it. Sequoyah proceeded to develop the only complete alphabet ever constructed by an individual. For this mammoth achievement, which resulted in rapid literacy among the Cherokee people, Sequoyah's name was attached to the genus of the most massive trees in the world, the giant redwoods—trees whose own leaves are merely tight, overlapping scales.

The metaphor of the speaking leaf is a powerful one. Yet in our quickness to adopt comparisons, we sometimes forget to honor the original image upon which clever metaphors are built. While ours is a culture dramatically affected by printed pages every day—prattling pixels on our computer screens notwithstanding—we spend relatively little time attending to the actual objects: real leaves. Oh, we rake them, burn them, and mulch them in the fall; stand in their shade in torrid summertime; and watch our philodendrons twine around our windowframes, if we remember to water them. But how often do we go deliberately out-of-doors, especially to listen to the leaves speak?

My odd homeplace, rich as it is in botanical contradiction and the happenstance of growth, differs little from any other vegetated zone in the complexity of its conversation. The textures, flavors of green, progressions of season (those hornbeams will be October gold, those oaks, flagrant red rags), are only the accents. The leaves speak in the dialects of warblers, the whispers and growls of the wind, the minings, riddlings, and stridulation of insects, and the chemistry of their own compact with sun, soil, and water. The point is that wherever leaf comes from bud, grows, falls, and goes to ground, the colloquy is endless and endlessly nuanced; yet seldom really heard.

Sequoyah's achievement was indeed large, and the dedication of redwoods in his name suitably proportioned—though he never saw such trees. Nor did his honor reach worldwide: in Britain, *Sequoiadendron gigantea* trees are known as Wellingtonias, sharing the honor of being named after Lord Wellington with rubber boots. But I doubt Sequoyah would care. Besides inventing an alphabet, he knew a language that few of the owners of the new talking leaves could hear. Each of us could strive to learn that lingo, could go forth among the silent plants, to listen, to hear what we will, and to learn from the old talking leaves. Only then can the black-marked pages of books, of magazines, of this journal, find their fuller meaning.

Autumn 1997

In the Eyes of . . .

Yesterday, at the old brick-bound pond that now serves as the compost pit, I met two slugs. One, a big, splotchy banana, was suitably arrayed on a banana peel. The other, a European brown, browsed a corncob. While the two imposing animals made an impressive molluskan tableau, I had to admit that I took more pleasure in the *Ariolimax columbiana* than I did in *Arion ater*. The olive banana slug is the more attractive of the two, and an important native species; while the brown slug, stubby and dull, is an alien species that represents a serious challenge to maritime gardens like ours.

That night, when I spotted a six-inch leopard slug at the cat's dish, my easy distinction between "good slugs" and "bad slugs" became conflicted: *Limax maximus* is another introduced slug, a serious competitor for pansies or lettuce, potted plants and potatoes. But it is also wonderfully handsome, long, sleek, slate-on-gray spotted and striped like our silver tabby, Virga. Too, this species exhibits a fabulous pairing: the hermaphroditic creatures dangle from a slime strand sometimes ten or twelve feet long to copulate and trade sperms. Once I begged Thea to let me watch an enamored pair copulate to completion before being frozen for the compost along with the gallons of browns and pints of little milkies. But then I decided to rescue them to watch their post-coital behavior, and they later escaped from the terrarium. Their kind has been much more numerous ever since. My mercy—or curiosity—had an unpopular outcome, despite the undeniable comeliness of the leopard slug.

So where lies beauty in nature? Clearly, in a place both focused and refracted by our own biases. Most of us would admire an alpine meadow rife with wildflowers or a resplendent quetzal. But do we all see what George Orwell saw in the common toad, or what I see in a slug? I know very well otherwise, just as I don't find each slug as appealing as the next. We all know the "Eew, gross!" emitted by teenagers who can't grasp their teacher's delight in a particular insect or slime mold. It seems we discriminate wildly in our attraction to the natural world and its elements.

Two experiences abroad underlined with permanent ink just how relative human fascination can be. Both incidents arose from field trips during meetings of IUCN (the International Union for Conservation of Nature). The first was a general assembly held in Ashkhabad, the colorful capital of Turkmenia. I was confronted in the hotel lobby by General Abrahim Joffe, a major figure in the Six-Day War and head of the Israeli Nature Reserves Authority. A large, gruff man, the general had just come back from a cruise on the Kara Kum Canal, a desert aqueduct that flows toward the Caspian Sea. "How was it?" I asked.

"All shit and frustration!" the general fumed. "Bad Russian beer, stuffy boat, and nothing to see but camels and endless reeds!" I told him about my trip to the Repetek Desert Reserve. We were to have sought the great gerbil on camelback, but plied the dunes in Soviet half-tracks instead, and never glimpsed the rare endemic rodent. "And tomorrow?" he asked. The Kara Kum Canal, I told him. "Ah!" he burst out. "It is a wonderful trip! You will love it."

And in fact, I did. General Joffe was right about the beer. But squeezed onto the bow with a Scot and an Egyptian, I found the Asian air off the water refreshing; the camels, coming down to drink between mountains of cotton, completely novel. The tall reeds indeed blocked much of the view, but they also supported a constant cross-channel traffic of brilliant blue-cheeked bee-eaters. And at lunch, in a native stand of tamarisk (a thirsty exotic in the American Southwest), I enjoyed my first face-to-face Asian monarch butterfly, *Danaus chrysippus*. What was "all shit and frustration" to the general had been a marvelous day out for me.

The other eye opener took place in Kenya, following a meeting at Tsavo National Park. Touring parks and reserves afterward, our vanload called at Samburu National Park to try to see leopards, reticulated giraffes, and Grevy's zebras. As well as the great naturalists Sir Peter Scott and Kai Curry-Lindahl, our contingent included two British wildlife officials who had been largely responsible for the inclusion of the rare, slender-striped zebra on CITES (the Convention on International Trade in Endangered Species). Naturally, Jane and John wanted to see "their" zebra in the wild. The leopard and the giraffe obliged, along with bustards, rollers, and gerenooks. But the much-sought equid remained aloof. Finally, in a tropical rain like an overturned bath, there stood our quarry—a single Grevy's zebra, abject in the deluge.

A few days later, passing back south through the Great Rift Valley, we stopped for gas. The Pope had been to Nairobi. People from all over Kenya had flocked to see him and were returning home. John Rudge, one of the Brits, was pumping petrol when he was descended upon by a large flapping nun in an excited state. "Oh!" she cried, hoping to share her rapture with a fellow European. "Have you been to see the Holy Father too?"

"Why, no," said John, equally excited by his recent audience. "We've been to see Grevy's zebra!" The sister's mouth dropped in absolute incomprehension.

Around here, a rabies scare has sent the populace scurrying for shots if they so much as spot a bat in their bedroom. As I netted a pretty little pipistrelle in my study today and turned it back out, I smiled wryly: how quickly caution becomes paranoia, and ignorance, revulsion. Of course, the consequences for bats have not been funny.

As members of the community of life, perhaps it should be our goal to find nothing in nature ugly; to revere the basic beauty of each living thing; to erect no hierarchies in our appreciation. But we are not made like that. When it comes to landscapes, plants, and animals, and certainly one another, we are creatures of strong preference. We are the beholders, and our eyes are ever crossed by our personal aesthetic, bias, fears, and favorites. It is good to recognize this. And perhaps it is enough always to try to expand our range of tolerance for the "ugly," even unto admiration; to seek the elegance of fitness in all things; and to realize finally, as Darwin said, that the tangled bank is clothed with "endless forms, most beautiful and most wonderful."

Wintrer 1997/98

Of Mice and Monarchs

So often we go looking for something and find something else: people in the street, books on a shelf, words in a dictionary. It happens in nature all the time, if we are open to what's out there. In fact, when people ask me how one becomes a naturalist, I say that being open to what's out there is at least as important as knowing what is out there, because the world can't help being an allusive, referential, sidetracked kind of a place. So it was with the mouse and the monarch.

I spent the fall of 1996 following the western migration of monarch butterflies. I wanted to see how these astonishing insects live during their great winter homecoming, where they go, and what it would be like to be guided by another organism, for months, and for many, many miles. Throughout my long journey—Canada to Mexico and back through California—I tagged every monarch I could, applying to the forewing a little adhesive label imprinted with a serial number and an address or phone number for reporting finds. Tagging yields much of what we know about monarch movements, though fewer than one tag in a thousand is recovered.

None of my monarchs was found, so I continued tagging the autumn emigrants the next year. One September evening I rolled across the Bickleton Hills from the Yakama Indian Nation down to the Columbia Gorge and chose a campsite on a bluff above the Great River of the West. The infamous gorge winds forced me to sleep in my little car—a frequent accommodation on quick field trips to awkward places. But a paddle-wheel tour boat, lit up like a dream from Samuel Clemens's youth, lulled me to sleep.

For a while, anyway. When a sound I took for rain on the roof began to resemble a race track heard from a distance, I realized I was sharing my quarters. Anyone who has ever camped out of doors, in a rustic cabin, or in an old Honda has likely become personally acquainted with *Peromyscus maniculatus*: the most successful mammal on the continent. Plenty of deer mice live in our old Swedish homestead, and I wouldn't have sweated this

15

one, but I had a heavy field day ahead and every time I passed into the precincts of sleep it did its little wakey-wakey walk over my hands, thighs, or brow, crinkled paper, or diddled my food.

I set a trap. I rigged my big butterfly net over the seat and laid a bait of cheddar. One, maybe two minutes, and it was there: black BB eyes catching the scant starlight, whiskers flicking over the mouse-bait cliché. The first time I snapped the net down the mouse escaped through the tangle of steering wheel, net handle, and hands. But it came back! And this time, I got it. I carried it out into the wind a hundred feet or so to the precipice. I considered launching it on a peromyscine parabola out over the gorge, penalty for sleep lost, but in the end I just released it into the basalt scree. The vague persimmon glow in the east looked entirely too much like dawn. But in the blissful solitude, I settled in and hoped for a hard hour's sleep, possibly two. Actually, only five minutes passed before the gallop of four little feet resumed.

On my next trip, in a milkweed swale a little east of Mouse Bluff, my field partner, David Branch, netted a deep-cinnamon female monarch. I gave her tag #09727 and sent her on her way, out across the broad Columbia in a stiff westerly. One month later, a young teenager named Jeremy Lovenfosse rescued an injured monarch from a road along Monterey Bay and took it home. He noticed the tag on its forewing and heeded the instructions to call in its serial number: 09727. The first native tagged monarch ever recovered from Washington State, she helped show how coastal over-winterers can indeed come from far away. In our efforts to conserve the threatened phenomenon of the migratory monarch, we need to know such patterns. Ms. 727's flight path describes a delicious route in my imagination. This one butterfly of passage on her "wings of flame, rising to the sun," in Jo Brewer's words, made my own journey whole.

But I sometimes wonder who made the more remarkable trek: the revered and lucky butterfly on her 700-mile *Wunder Fahren*; or that one, utterly irritating rodent determined to keep me company on the road? Traveling with mice or monarchs as our guides, we can expect to be joyfully sidetracked, returning again and again to that rich territory where place, task, and company conjoin.

Spring 1998

Turning Fifty on Silver Star

When Thea and I married, we built the living-room ceremony around friends, family, flowers; autumn leaves, the local judge, a little Walt Whitman; and a best-forgotten sonnet with a well-remembered message: that getting OUT would define our lives together, so help us. And gotten out we have; but never enough.

A regrettable but formidable tendency ensures that many would-be naturalists betake themselves out-of-doors all too rarely. This misdirection not-so-subtly subverts the very impulse that drives us to work on behalf of the outside world. So biologists spend their time in labs, committee rooms, and conference halls, while conservationists haunt offices, meetings, and legislative chambers. Rangers administrate, managers delegate. And they all devote much more time, unless they are clever or vigilant, to a computer terminal than to the wild, be it yard, park, or mountain fastness.

Years ago, when I worked for international conservation groups, most of my colleagues had been inspired by a love of nature. But they had grown far from the model I envisioned for myself: an engaged, yet oft-sauntering naturalist. Plainly, they never got out; and then they forgot.

I took steps to avoid that fate, and over the years I have made certain to get out often, if only modestly. That is one reason I live where I do, where even fetching the mail is an adventure. But I have not escaped that insidious counter-pull altogether, especially when it comes to longer, more physically demanding excursions far from the desk, mail, telephone, terminal, and all other anchors on the ambler's drift.

In our early twenties, Thea and I belonged to the University of Washington Conservation Education and Action Council. We watched David Brower movies on the North Cascades and other imperiled wildlands, secure in the knowledge that we would see all those charmed scenes with the aid of boots and backpacks. Thirty years later, the treks to high meadows and wilderness beaches have been far fewer than we'd anticipated. The summer we both turned fifty, we decided to mend our ways.

I suppose we did it partly to prove we still could, partly out of a sense of summers slipping by rapidly and irretrievably. No matter. On July 18 we girded, loaded, and stretched our loins, and set out on a trailhead recently redeemed from an old forest lookout road in Washington's southwesternmost Cascades. Silver Star Mountain was reputed to be a fine place for butterflies. Since these insects favor flowers, the mountain held high promise for Thea's primary interest too.

Our first morning broke to Mount Adams emerging from the night into a slurry of rose mist and summer sun. The snowy cone suffered in competition with the dawn-and-dew-struck flora. Tall orange Columbia lilies swayed on a light breeze all around the carefully placed gray dome of our tent. A canopy of creamy umbels and full-maned yellow composites wrapped outbursts of magenta paintbrush, furry mariposa lilies, blue gentians, and scarlet columbines. Our perch on the flower-strewn slope was solitary, since a lack of surface water deterred backpackers. You wouldn't know it from the lushness of the turf. For two days we explored the slopes, trails, hollows, and peaks of Silver Star, and I turned over my first half-century in the embrace of four white volcanoes.

And the butterflies? I have never seen such a spectacle in butterfly-subtle western Washington. Only fourteen species, but thousands of golden western sulphurs, hundreds of wax-and-cherry Clodius parnassians, chalcedony and Edith's checkerspots by the score. Sharp-eyed Thea spotted the small brown chrysalis of a western meadow fritillary on a granite chip beneath its host-plant violets. And as sunset glazed Mount St. Helens, she saw three backlit butterflies bed down: a diaphanous parnassian on a pinkened umbel, a meadow fritillary on a rush, and a checkerspot on a blown dandelion. There they stayed till morning rays warmed their wings and sent us down the mountain to water.

One week later, we again took up our beast-of-burden packs for a trek into the proposed wilderness of the Dark Divide, northeast of Silver Star. On the summits of Sunrise and Jumbo peaks, among native plant society friends and mountain heather, we found alpine butterflies never before recorded in Skamania County. Come September, we kept Thea's fiftieth among the red-and-blue blaze of ripe huckleberries that gave Indian Heaven

Wilderness its name. If our loads were not yet second nature, neither was the experience secondhand. I don't doubt that next summer will see our packs further exercised, our canoe wetted more than it has been lately. We are rebuilding the habit of getting out, often and well.

You don't have to backpack, of course, to taste the "tonic" that Thoreau distilled from his walking. But as we age and our lives complexify more and more, we must ardently resist the busy demon that would keep us in. It doesn't do to become prisoners of our own commitment. We deserve to experience what we labor to preserve. For the good places' sake—and for our own sake—we are wise to get out, as often as we possibly can. After all, going afield, as the old naturalists called it, is its own reward.

Summer 1998

Roll Call

Haying is going on right now where I live, each farmer with one eye on the rows, one on the clouds. Here in Willapa, people talk a lot about grass—also Douglas-fir, black-tailed deer, and elk; coho salmon, sturgeon, and Dungeness crab; Holsteins, Herefords, and slugs. In the country, many people (though fewer each year) still take their livelihood directly from other forms of life. Townspeople are less likely to connect with nature on a regular basis. Some, such as bird or butterfly watchers, wildflower and mushroom fanciers, organize their free time around nonhuman encounters. But such folk are uncommon overall, and considered strange by many of their neighbors: eccentric, obsessed, if harmless. The majority, in fact, shows little awareness of other life forms beyond cats, dogs, lawns, and fellow humans.

To resist this all-too-easy ignorance, I recommend making a written accounting of the many species we brush up against daily. For example, a spate of recent travels, and the face-changes that spring brings, yielded these contacts for me:

That damn dog who barked me awake. A Wilson's warbler incessantly seeking a mate. Two *Felis catus* patted goodbye, each made (in part) of six or eight species of recycled rainforest rodents. The badly broken coyote limping along the road on my way to the airport, where tall cottonwoods leafed out limbless in the "designated wetland," survivors of last winter's ice storm. The sweet balsam scent of their unfurling buds.

Then, in Fairbanks, exploding aspen catkins; sweet birch sap flowing from beaver cuts; waffles sweetened with three species of *Vaccinium;* and last year's tart lingonberries lingering on the forest floor, sodden with snowmelt. Browsing moose and musk ox cows and calves, nesting tundra swans, and sandhill cranes flying over the muskeg. Early purple pasqueflowers blooming under black spruce on the University of Alaska campus, a benignant brown bear looming behind the museum podium, and taxidermically malevolent polar bears presiding over each Alaskan airport lobby.

Back home again, violet-green swallows nesting by my study window, tree swallows on the porch, barn swallows in the canoe room. Big bird biomass at smelt by the covered bridge—eagle, osprey, gull, merganser, cormorant, corvids. Cattle in deep green grass, anise swallowtail and mylitta crescent over phlox. Thousands of dispersing Asian ladybirds, one or two of them rounding my computer screen and glasses rims at all times. Rug-munching clothes moths, a red ant biting me where it never should have been, and tiny ants invading the bathroom.

A giant carpenter ant on the road with me as horizontal Sitka spruce, western hemlock, and western red cedar pass by on log trucks, while California poppies and Scots broom splash I-5's shoulders with bright color. Aloft again with Chief Flight Attendant Diane Alder, Pilot Scott Redheifer, and eight recognizable species in the lunch.

Life is a litany of other species. So why are we often oblivious to all but our own? Earlier cultures had a basic, intensive, and entirely essential working knowledge of other species in their midst. In my grandmother's day, the leading of botany and bird walks was a routine part of the grammar school curriculum. Yet in Fairbanks, where many people were studying "nature," not a single person I met had noticed the outrageous early eruption of purple pasqueflowers.

On another recent trip, I traveled by that species of transport named for a fleet grey dog. To escape the clouds of *Nicotiana* vapor rising from the riders' spontaneous combustion at each stop, I took short, quick walks to see what I could find behind the run-down depots: milkweed and sweet clover in a vacant lot; *Gaillardia* and petunia in a planter; hops, yeast, and barley in a pint glass. I never saw another passenger leave the stations. As I stretched my seat-sore muscles, I lamented that what seemed absolutely natural, even essential, to me drew only puzzled glances from my fellow travelers.

The species roll call makes an illuminating school assignment or journal exercise—not necessarily going forth to "document biodiversity," but simply taking into account the other members of what poet Pattiann Rogers calls "the family." You don't have to know everyone's name; you can list "the purple flower with the woolly leaves by the corner," or "the bird that squeaks like a rusty gate each morning." Once a neighbor is accounted for, the next step is knowing who it is, what it does, and how it fits in among the wider community of life.

Gary Snyder reminds us that "one is in constant engagement with countless plants and animals ... the non-human members of the local ecological community." When we fail to pay attention to our evolutionary neighbors, we deprive ourselves of much of the pleasure, comfort, fascination, and companionability of the world. Taking roll, we begin to sense our own fit.

Autumn 1998

Old Growth

It was the title of the thing that grabbed me first: "Biotic Aspection in the Coast Range Mountains of Northwest Oregon." An "aspection" turned out to be an overall "look around" at the forest's lifecycle from all sides— organisms, seasons, soils, and weather. The 1951 paper described James Macnab's pioneering 1930s study with his students from Linfield College. Working with primitive equipment and teaching themselves as they went, Macnab's team painted a picture of place in rare detail. That place was Saddleback Mountain, a misty hill poking out of the old growth only a few miles from the Pacific.

Jane Claire Dirks-Edmunds, Macnab's chief field assistant, went on to earn her doctorate, take over the study, and become a respected ecologist in her own right. As a lively and venerable retired teacher in McMinnville, Oregon, Jane Claire has written a rich memoir of this classic ecological study. Her book, *Not Just Trees* (published in 1999 by Washington State University Press), crystallizes much of what was learned on Saddleback Mountain. This is good, for the forest they plumbed—to a depth seldom attempted today—is no more. What should have been jealously saved as a priceless baseline for the maritime Northwest rainforest has since become an ordinary, short-rotation industrial woodland. The one stroke of grace is that the forest's minstrel lived to tell its rich, sad story. If *Not Just Trees* no longer matches an actual place, it tells us what a real forest could be if only we would let it, and models a comprehensive approach to the infinite complexity of old growth. Everyone involved with forest management and protection should heed this wise and humble tale.

Listening to our elders is no new idea, but we can never remind ourselves often enough. We need only think of late, great conservationists such as Aldo Leopold, Rachel Carson, William O. Douglas, and the recently departed "Mother of the Everglades," Marjorie Stoneman Douglas, to remember how their advice and examples empowered educators and activists seeking to confront the ecological challenges of their time. Our living elders—Margaret Murie, Victor Scheffer, John Hay, and Hazel Wolf, to name a few—have extraordinary gifts to give, if we will hear them out.

We should hearken especially to our naturalists, those who know the living details of the world, how the land once was, and how we ought to address what's left of it. This past summer was a tough one on seniors of this species. Within a few weeks we lost Ron Taylor, one of the finest field botanists in the West; John Hinchliff, the beloved and devoted keeper of Northwest butterfly data; and John Burroughs Medal-winning writer and the dolphins' best friend, Ken Norris, one of the last two holders of the title "Professor of Natural History" in the University of California system. Though each left behind the heart of his knowledge in memorable books and with those he taught, we all were diminished when they departed. What we lose when elders pass is personal wisdom based upon lives well lived, unique knowledge of former times and places that will not come again, and, in the case of naturalists, ways of seeing that we could dearly use as we seek to understand, to adapt, to reform, to restore.

Few of us have escaped the bitter remorse that comes from waiting too long to see an aged relative, an afflicted friend, a failing mentor. For years I tried to find Ben Leighton, who had dropped out of sight soon after his *Butterflies of Washington* appeared in 1946. A serpentine trail, spiked with dead ends and coincidence, finally led to a convalescent center not two blocks from my stepchildren's home! I was eager to show Ben modern works based on his beginnings, to tell how his contribution still mattered, and to ask questions that only he could illuminate. His nurse said he was strong, and suggested that I "skip the storms on the pass and come in the springtime." But shortly after New Year's, Ben Leighton died.

So often, we don't seize the opportunity to ask the questions we need to ask. But, if we could anticipate the true value of the answers, we'd know we ought to ask them early and often—and listen up. For just as we depend upon the old-growth trees to speak "the forgotten language" of the forest (as poet W. S. Merwin puts it), so do we need our own elders to show us how things are, were, and ever shall be. Few of our old growth will leave books behind to speak for them, as Jane Claire happily has. It is never too soon to go humbly among these thinning groves, with open ears and eyes. But blink, and it may be too late.

Winter 1998/99

NOTE: *Jane Claire Dirks-Edmunds passed in 2003 at 91; Hazel Wolf in 2000 at 101; Mardy Murie in 2003 at 101; John Hay in 2011 at 95, and Vic Scheffer in 2011 at 104.*

The Element of Surprise

The sweaty satisfaction of activism has a dark flip-side. Its name is burnout. Which of us, worn down by too much devotion to an enterprise that needs us, has not known the distress and guilt that bailing brings?

Yet I believe nothing freshens the wrinkled will like immersion in the natural world. This is especially so for conservationists, who way too often forget what they are working for, what it looks and smells like. And nothing restores the wonderment like sheer stupefaction: the shockingly novel sensation that awaits every watcher who goes forth to indulge in the blessed ordinary, then rediscovers that it seldom is. This antidote is free to anyone willing to attend the infinitely generous offerings of happenstance.

Take, for example, bats. Naturalists know bats, though commonly reviled, to be valuable and fascinating animals. We go to see their phenomenal fly-outs, in black-cloud millions, from places like Bracken Cave in Texas or Austin's Congress Street Bridge. But a single bat where you least expect it can be just as stunning.

After Christmas snow and a month that brought a yard of rain, an early January day came bright and balmy. Snowdrops swelled near to bursting in the dooryard; the long catkins of hazel and the first skunk cabbage, by their golds, signaled the halting but sure start of spring. Indoor work had gone stale, and we couldn't resist a walk.

As Thea and I climbed a grade above Gray's River, looking down valley, we spotted what appeared to be a bird, then, we thought, a big red moth. But it didn't fly like a song sparrow, and we were months away from the cherry-and-mahogany ceanothus silk moths, which, in any case, don't fly by day. Well, neither do bats, which should have migrated or been hibernating by now. Big and little brown bats appear in our bedroom many a summer night, and hunt insects over the yard; but in January? Then I recalled that one year before I'd seen a bat working this very same stretch of road, just after sunset, as a barn owl and a tree frog called on the hillside.

Our subject fluttered back and forth past our faces, its own ears and face readily visible. Against the blue sky of the east, its wing-strokes showed all above its back, like those of a short-eared owl. Then, backlit by the dropping sun while hawking midges at the meadow end of its hundred-yard circuit, its thin membranes that do for wings flashed bright red. Darting back again, under maples, alders, and spruce, it looked for all the world like a big brown butterfly.

That perception took me back to the Colorado River, when I was following the migration of monarch butterflies two years before. Gary Nabhan had alerted me to monarchs moving through the Grand Canyon earlier that fall. When he first saw them coursing above his raft, he'd taken them for bats abroad in the daylight. I chuckled at that until I stood on Navajo Bridge, looked far down into the canyon, and beheld the desired monarchs milling in the late sunshine. Surprise: they were bats.

Our Willapa winter bat flickered in the sun, veered, and motored past again, its warm brown dance reminding me of another false call. Following the Orion Society's Forgotten Language Tour in Texas last autumn, I canoed with local conservationists working to save a wild bayou near the mouth of the Trinity River above Galveston Bay. We'd slowly paddled among alleys of fluted cypress buttresses and their gnomic knees, dodged under great golden orb-weaving spiders on their ten-foot webs, spooked little blue herons and roseate spoonbills. Then, in a narrow, leafy channel, a powerful brown bat appeared down bayou. We stared as it zigzagged across the black water; but something wasn't right, and when it alighted on a hardwood bough, I yipped, "It's a black witch!"

"A *what?*" came back four voices. It was indeed the huge moth, beautifully striated with black and turquoise, a southern species that sometimes emigrates north. I had only seen *Erebus odora* once before, twenty years before, flapping its way through the hot canyons of downtown Denver.

Surprise! It keeps the world fresh, no matter where you find it—the percussion in the ceiling that announces the presence of woodrats; the vole in the cellar, deep in deer mouse territory; the salamander in the sump-pump pool. The surprise that keeps life lively can be as great as rising this morning to behold twenty-four Roosevelt elk leaping a fence beside the covered bridge, until the youngest calf discovers the open gate. Or as small as an optimistic bat, spinning through the January sun.

Spring 1999

A Declaration of Independents

I don't know how many times I've heard someone exclaim, upon seeing a butterfly, "Doesn't that prove that there has to be a god, to put such beauty into the world?" The countervailing view seems just about as logical to me: how could there be a caring god in a world with death and sinuses, let alone leaf blowers? Silly as such ipso facto arguments sound, I love the one proclaimed by Benjamin Franklin: "Beer is proof that God loves us and wants us to be happy."

I write in an agreeable Oregon brewpub, a pint of good India Pale Ale before me, old Grateful Dead playing soft on the PA. Thea and I have just seen *You've Got Mail* at the old single-screen, main-street movie theater across the way, for $2. I am trying to decide how I feel about a film that tackles a serious issue, the destruction of independent bookshops by giant chain stores, then blurs the issue with a happy ending.

As an author, one of a class of people whose hard-earned income (royalties) is generally halved by sales of their products (books) in the superstores (those discounts come from somewhere), I have a stake in the preservation of the literary marketplace's diversity. I also happen to love a good "indie," and (along with many of my writer friends) will not sign books in a chain. Many of us also enjoy a good ale, and as members of a bioregional community, we consider good pubs and good bookstores to be elements of local diversity. (In England, one's neighborhood pub is actually referred to as "the local.")

Now, alcohol is no joking matter, though it is the subject of almost as many quips and gibes as sex: "Beer—no longer just a breakfast drink," reads a placard over the bar in my favorite alehouse, where the publican will serve nothing advertised on television. But at a time when many people are seeking to defeat their dependencies, and the evils of strong drink in homes and on the roads ruin more and more lives, beer is a serious subject indeed. This is no paean to drink or to drunkenness.

Still, Ben Franklin had a point. Properly approached, beer has its beauties—not the least of which is the inspired symbiosis among several organisms that its decoction entails: yeast, barley, and hops, mixed with pure water and time. The fermentation of malted *Hordeum distichum,* through the sugar-converting properties of *Saccharomyces cerevisiae,* fixed by the essential oils of the noble *Humulus lupulus,* creates a spectrum of rich colors and flavors. Malt beverages range in hue from fresh straw through fall foxplume to coal-scuttle black, in taste from dry-bitter to sweet-fruity to malty and chocolaty espresso, involving most of the talents of the tongue and all of its surface. In addition to all this, beer offers up an adventure in natural diversity.

At one time, each valley in England had its distinctive ales, just as every Scottish glen distilled its own malt whiskey, with the smoky tincture of peat instead of the bite of hops. Using different yeasts and recipes, Continental brewers created lagers and pilsners as numerous and distinctive as the bitters and stouts of the British Isles. This Saxon beer heritage dominated in the U.S. until one of the great extinction events of our time erased it. Prohibition extirpated local tastes, traditions, jobs, and literally thousands of strains of yeast—individually selected forms of life that will never live again. Once down the drain, their DNA was as gone as that of the Xerces blue butterfly or the great auk. And when beer came back after The Long Thirst, it was in a monolithic fountain of yellowish, watery lager lacking character or distinction.

The same sort of pogrom threatened British beer in the 1970s, when six giant companies took over many regional and local breweries. A discriminating public bit back through CAMRA, the Campaign for Real Ale, saving many of the family and village breweries and spurring the first new ones in a century. This success inspired the microbrewing movement in North America, with its center of diversity smack in my biome of the Maritime northwest. Which brings us back to this pub, this pint, and *You've Got Mail,* because the Pacific Northwest is also famous for its bookstores. Yet, even as small breweries have multiplied, our beloved neighborhood bookshops have been gobbled up by the ignoble book barns like cashews at a thirsty bar.

Now, big is not always bad. Look at Yosemite, or Powell's Books in Portland. What one deplores is when BIG rolls over the small, good things, sanding down the different as it goes. As for *You've Got Mail,* I still haven't figured it out. Maybe if Tom Hanks and Meg Ryan weren't so blasted cute. I like to think there will be trouble ahead for their romance unless he gets a different job. But I feel confident in saying that readers commit a progressive political act on behalf of community when they shop in an independent bookstore. And we who drink beer—pray, in moderation, and at all due risk to our waistlines—do the same when we choose the local, the real, and the good. This excellent IPA, in a revivified country hostelry with clean air and a lively clientele, is the proof—if not of a god who loves us, at least of a truth we need to repeat over and over: that celebrating and embracing the local can sometimes save it.

Summer 1999

Naming Names

Stuffed into the uttermost bowel of a 757, at the ragged end of a red-eye from Portland to Baltimore, I am consoled by visions of the Siskiyous. Hours earlier, my wife, Thea, and I had returned from those serpentine-girded mountains in southwestern Oregon. We'd long desired to hike, botanize, and butterfly in that wild country, and now its delights bloom anew in my mind's eye, coaxed forth by a gentle litany of names:

Kalmiopsis. Azalea. Ceanothus. Brodeaia. Wavy soap plant. *Sedum laxum.* Gorgon copper. Leanira checkerspot. Babyfoot Lake. Eight Dollar Mountain. *Papilio indra. Fontinalis.* Golden chinquapin.

Lost among the sensations fetched back by these names, I think back also to a question put by a student in a recent writing workshop in the North Cascades. "Don't names just get in the way?" she asked. "It seems that by classifying plants and animals we just objectify them. Shouldn't their beauty be enough?" I have heard many versions of this resistance to the practice of naming names. I first encountered such an attitude when I was in college, in the midst of an excited learning frenzy, gleaning the names of everything I could. Returning to Colorado on a spring break, I found that one of my best friends, who was just discovering hiking and nature, thought that naming detracted from the root enjoyment of flowers and creatures.

While for some this is doubtless an earnest truth, I believe that for others it represents a rationalization. From many conversations, I have discovered three common reasons for avoiding Adam's task: First, many observers are simply intimidated by the sheer number of things with names out there. They feel they can't possibly catch up, so why try? Second, some people are embarrassed that they know so few organisms by name—though this isn't surprising, since it has been generations now since nature study was considered a standard part of primary education. Third, a lot of would-be naturalists are just lazy. As an essentially lazy person, I feel a certain kinship with these last folks. They know it will take some work and application to learn some species by name and cop out instead with the impressionist's excuse.

But there are at least as many good reasons not to shun the names of the elements of life. For one, by knowing the identities of other living things around you, you come to understand their relationships much better. This allows you to appreciate your own evolutionary heritage, and how each organism fits into the lineage. For another, these are times when the need for understanding, documenting, and monitoring biological diversity has never been greater. We can all take part in this. But there is simply no way to account for life on earth without learning to recognize the constituent parts and applying names to them. Just as surely, observant and curious people find their pleasures afield vastly enhanced by intimate acquaintance with more and more species. Just watch how birders, butterfliers, wildflower watchers, and other seekers ramp up their enjoyment out of doors by making, and naming, new friends wherever they go.

None of this is to discredit those who truly wish to take nature at face value; nameless does not mean faceless, and keen pleasure and communion are possible without knowing the nomenclature. After all, there was no good field guide to Siskiyou plants, and Thea and I did not feel like carrying two or three large floras and spending all our precious time keying. So for many plants, we went generic: that frilly campion reminiscent of the English ragged robin, those elegant little two-toned violets. But at least we often knew the genera or families, could place their relations. And that would be a worthwhile objective for many of us, and a reasonable one. You may not know a cardinal from a pyrrholoxia or a blue jay from a scrub, but with a little watchful time in the open spent with your Roger Tory Peterson, you can easily come to know that this bird is a tanager, that a flycatcher, and the other a finch.

We are landing in Baltimore. When I get to the Orion Society's Fire & Grit conference, all the humans will have nametags on. The other species will not, but that's what field guides are for. As soon as I hit the ground, I plan to scan the trees (sassafras, sycamore, shagbark), looking and listening for warblers (Blackburnian, parula, prothonotary). Along with the Carolina wrens and the great spangled fritillaries, the writers and the activists, they will be my neighbors for a few days.

And that's what it comes down to in the end: knowing your fellows in the neighborhood of living things. You can smile politely and fake it when you don't know the folks next door or down the block. But when you call them by name, recognition and relationship become possible. And aren't we more likely to *care* about neighbors we know by name? As the great nature writer Ann Zwinger perfectly put it in *The Nearsighted Naturalist*, entering new territory is "like walking into a big party where at least I know a few families, and recognize some friends." Those who take the trouble to identify plants and animals, she rightly says, are "at home in a natural world that will offer them challenge and pleasure the rest of their lives."

Autumn 1999

Down at the Grange

Walking down by the old covered bridge one day during my first fall in this valley, I met a dairy farmer named Bobby Larson. His family had pioneered the south side of the river, while the Sorensons, who sold me my place, farmed the north, and they'd built the bridge between them. Conversation settled on the Alaska Pipeline, and though our views differed, we established a rapport that has only warmed these twenty years since. Before we parted that day, Bobby invited me to attend a meeting of the local Grange.

I've been a member ever since. Friends are sometimes surprised that I belong to what they imagine to be an antiquated, geriatric, quasi-masonic, and reactionary old farmers' club. But that image is only partly accurate. Antiquated, yes; or at least venerable. The Patrons of Husbandry (Grange) was founded in 1867. Gray's River Grange #124 arose in 1902, its first Master being H. P Ahlberg, the Swedish immigrant who built my house. The organization was originally a protective society for farmers, who teamed up to resist railroads and other big business that was squeezing small producers. Grange ritual, once necessary for secrecy when members were persecuted like union organizers, is indeed vaguely masonic. But I found it to be inoffensive: the lovely rhetoric full of natural images, the ritual laced with symbols of the field, agricultural implements, and for the office of Gatekeeper, a pole topped with an owl.

True enough, Granges have closed by the bushel and meetings are largely gray-haired. But I found a vitality among the oldtimers, refreshed by younger newcomers like myself, eager to meet the locals and find our place in the deeply ingrained community. Thirsty for new blood, the Grange was welcoming, even of a post-urban, long-haired environmentalist who brought Wendell Berry poems to read aloud during the programs.

Not that Grange politics always match my own. As it grew and became top-heavy and complacent over time, the Grange got conservative, turgid, and closer to big agribusiness. I find many of its policies at the national and state levels distasteful. Grange tends to promote pesticides and private property rights, and oppose wilderness and dam removal. But at the *local* level, Granges have been engines for positive change, conservation, and the arts, and homes for true community connection. Ours, for example, secured public water—the best I've ever tasted—for a current total of 258 customers; spearheaded the restoration of our historic covered bridge (the only old one left in Washington); re-enacted Captain Robert Gray's arrival in 1792; opened a shop for local artisans; created a park with big pioneer elms and riparian habitat beside our once-busy tidal basin; and supported the establishment of nearby conservation areas for ancient forest remnants—the last old growth in the county—though the membership includes logging as well as farming families.

Tonight, as I write (I always take work to Grange, for the meetings can drag on), we discuss how to protect local utilities and a 911 system where the dispatchers know everyone. We also talk about deeper dredging of the shipping channel in the Columbia River, and the bullying tactics of the Army Corps of Engineers. The emphasis is on keeping the local local and the small small, and you'd be forgiven for thinking you'd dropped into a meeting of the E. F. Schumacher Society, or the Sierra Club, though good Grangers would probably never go there.

But what I love most is the stories. When Merlin tells of how his Uncle Norman, whom I remember peering barely over the steering wheel of his Ranchero wagon, once drove his tractor up the side of his big chestnut; when Carlton recounts how his mother Agnes—whose silver braids I recall as she served her last years behind the counter of the general store—took the payroll up to the camps on the logging railroads; or when the travails of particular cows, horses, or black sheep ancestors are passed around with the pie and coffee or the once-a-year oyster stew, I feel more and more woven into the history that I have chosen as my own.

Now, not every community has a surviving Grange. But this isn't really about Granges. It is about finding a place where the hidden heart of community still thrums, and becoming part of that place. It's about engaging with a social natural history. Too often, we who reinhabit the countryside find ourselves even more isolated than we had been in the city;

and those still in the city imagine that true community resides only in some idealized rural Brigadoon. The fact is that you can plug firmly into town or country by attending not only to the plants, animals, soils, and waters, but to the people who made those connections before you.

The magic meeting ground might be a village hall, a library, or a cafe; the post office, PTA, or pub. In one town, the Odd Fellows became the focus; in another, the restored schoolhouse. Wherever they spring up, such repositories of the old tales and lifeways are well worth seeking out for those willing to sit and listen for a spell. If a community can be said to have a soul, it can be found among the ghosts and the living, down at the old Grange hall.

Winter 1999/2000

NOTE: *If you would like to read more about Gray's River Grange #124 in recent years, including the remarkable reign of Nirvana co-founder and bass guitarist Krist Novoselic as Grange Master, you may refer to my book* Sky Time in Gray's River: Living For Keeps in a Forgotten Place *(Houghton Mifflin Harcourt) and the handsome Web site maintained by Brother Krist at http:// graysriver.grange.wahkiakum.info/grays_river_grange/ Here you may also find our joint poem/guitar composition, "Notes from the Edge of the Known World," an exploration of natural communities human and otherwise, or else hear it on You Tube.*

Of Cabbages and Queens

My wife, Thea, came home from a walk one day with a winter bouquet in hand. I recognized the ferny leaves and the spidery, complicated flowerheads, colored the greeny-white of elder flower, some spread flat like a yarrow head, others cupped into a basket, each with a deep purple spot in the center. "Queen Anne's lace!" I said. "Where did you find it so late?"

"Just down the road," Thea said. Obviously I hadn't been getting out enough.

Though I know better, I am always surprised by plants that seem to be blooming unseasonably. As a Colorado native, part of me still thinks of winter as a flowerless time. There, as across most of North America, Queen Anne's lace joins the panoply of winter weeds whose brown-netted skeletons lend the sere or sodden landscape a reminder of life. But here in the maritime Northwest, this delicate umbel, like the related but hulkier cow parsnip, grows and blooms late into the autumn and not uncommonly past the first frosts. Still, to find it yet green, with creamy inflorescences unspent, between Thanksgiving and Christmas, always feels like a gift. And that was even before high pressure settled over the region for the last week of the year, bringing such blue-and-gold days that yellow Welsh poppies, red kafir-lilies, and forget-me-nots the color of the abundant Steller's jays all wakened into bloom. Marah—what many call wild cucumber—started in again to sprout and twine and bud, months ahead of schedule.

Queen Anne's lace is not a native plant in North America. It arrived very early and spread rapidly through fields and vacant lots, along the lanes and roadside verges, so thoroughly that most people are surprised to hear it is an exotic weed. Folks look even more shocked to learn that Queen Anne's lace and the common carrot are the same species of plant. Cultivated carrot and wild carrot, the weedy, roadside type, both belong to the species *Daucus carota*. It's just that the escapee goes to flower, while carrots aren't generally allowed to; but examine a field of seed carrots, and you'll see the resemblance. As for the name, that elegant, finely filigreed double-umbel is reminiscent of the lacy collar worn by women of style in the time of Good Queen Anne, three *fins de siècle* ago. In folklore, the puffy white flowerhead is Her Majesty's great starched ruff, while the little heliotrope button in the center is said to be her bonnet. A botanist would say that the single dark flower in the center is probably ultraviolet reflective, a bright signal for attracting pollinating insects to the middle of the landing platform, like a glowing white X on a helipad.

Just as carrots and Queen Anne's lace are conspecific, cauliflower, broccoli, cabbage, Brussels sprouts, and kale are all selected forms of the crucifer collectively known as *Brassica oleracea*. These, too, are winter-tough plants. Visiting the city, we see the bright purple and ivory ornamental kales in beds where tulips will later dominate, and call them "fancy cabbages." At home, our Brussels sprouts produce right through the winter. This year, for the first time ever, I've found caterpillars of cabbage butterflies feeding on the sprout plants all winter long. As much as any other indications, organisms adapting to northern winters like this may announce our ameliorating climate (*sensu Jane Eyre*: " ... the frosts of the winter had ceased; its snows were melted, its cutting winds ameliorated").

In these first few months of the twenty-first century, everyone seems to be watching for beacons of change. Some seek signs of prophesy come true, convinced that arbitrarily fixed numbers (like 2000) somehow matter to the universe. Others, bored with the regularity of life, hope for something— ANYTHING—to happen, just to liven things up. An extraordinary thing actually did happen, for anyone who was watching. For the final solstice of the last century, the moon shone bigger and brighter than any of us have ever seen it before or ever will again, and the skies were clear here to see it, nearly a miracle in oft-beclouded Cascadia. As Thea and I walked down the road to watch giant Luna's moonglade on Gray's River beneath the pewtered outline of the covered bridge, a warm breeze blew from the east. A single great blue heron stood on the shingle below the bridge, set in solid silver, easily able to spear fish in the spotlight.

I am not immune to the desire for signs. We all long for clues from the universe. But not being much of one for numbers or prophets, I just watch for the return of the rain with the waning of the moon. For the freezes that will still come to shrivel the marah and send the slugs back into hiding; then the unfurling of skunk cabbage, and the breaking of salmonberry buds in scrubby brakes between town and forest; the sun climbing and sinking just the same. These are the signs that count for me: the signals of the seasons, the semaphore of unruly life in all its flow and flux. The earth spins, and we go on.

Spring 2000

When Things Go Wrong

Aldo Leopold may have been the first to articulate that special pain reserved for those with an ecological awareness, but many have discovered it since. You simply cannot be attuned to the finer details of the natural world without feeling tortured when the fabric of the land unravels around you. This is one of life's nastier trade-offs. Sure, it's possible to remain unmoved when a marsh is drained or a meadow paved: but only for those who have missed the meaning of meadow and marsh. If you want the pleasure of the fruited world, you must be prepared to take the anguish with it.

There is no way to reconcile yourself to this bargain, nor does it get any easier. We speak of "disaster fatigue" among viewers of modern TV news, and I suppose a kind of eco-fatigue may set in after your tenth spoiled landscape of the week. But the refraction period for such feelings is cruelly brief, especially in your own neighborhood. Just as wonder refreshes, the capacity to be dashed renews as well.

Certain high forests in central Mexico host overwintering monarch butterflies, hanging from tall firs in their brilliant millions. I have been fortunate to visit these magical sites many times over the past twenty years. Most recently, I took part in an Orion Society pilgrimage to Michoácan to honor poet-activist Homero Aridjis. It was our good luck to visit the monarchs in the company of biologist Lincoln Brower, whose decades of research on the migratory butterflies motivated a Mexican presidential decree that created sanctuaries at several of the winter sites. At our first destination, Sierra Chincua, the butterflies (mobile from year to year) clustered near Brower's original research camp. As always, the vast gathering of cinnamon sailors stunned the senses; their soft sussurus on the air lulled me into the sense of grace one often feels among them. But this time with the monarchs, something was badly wrong.

Against Lincoln's advice, local officials opened Sierra Chincua to visitors so that the traditional landholders might benefit from the new largesse that monarch tourism is bringing to this very poor region. Indeed, ecotourism offers an economic alternative to cutting the high, fragile forest of oyamel firs. But this gamble is turning out to have serious side effects, with the potential for even bigger losses. Thousands made the high-elevation trek that Saturday, many on horseback, wearing bandannas and masks against the rising cumuli of thick brown dust. The *caballos* and *caballeros* cut deeper and wider trails that braided over the stony slope, effacing the lupines and salvias that used to swaddle the thin soils. The scent of crushed yerba buena rode over the dust, as monarchs suffocated, their spiracles clogged with the atomized soil of Barranca Honda. I was shocked by the deterioration that had occurred here since just the previous year.

Still, the forest's lives unfurled about us. Clown-faced red warblers hunted insects among bush senecios. Monarchs that hadn't fattened enough for winter probed scarlet salvias for scarce nectar. As the sun plunged under an upstart cloud, we witnessed the phenomenon known as a "cloud bomb," whereby baskets of butterflies adorning the boughs take to the sky in a sudden explosion of orange.

The next day, we visited El Rosario, the original monarch mecca for tourists. With some guides, guards, fences, and signs, Rosario is better prepared for the onslaught of enthusiasts. But the villagers' shops and facilities are expanding higgledy-piggledy, and each year their fields creep farther up the slope where butterfly trees recently grew. In the nearby towns of Ocampo and Angangueo, new sawmills mock the government ban on logging in the sanctuaries. For the first time, the losses overcame my delight in being among the monarchs.

And when I got home, I found our own valley under attack. The old farms that comprise the heart of this rural enclave were recently broken up, and much of the land acquired by an entrepreneur from outside. After assuring the neighbors that he wanted to farm and would never develop, he announced plans to plunk down a dense subdivision of manufactured homes on prime farmland near the only historic covered bridge in Washington, which valley residents worked hard to restore just a few years ago. Besides raising serious concerns about flooding, septic drainage, and wildlife, this scheme would destroy the bucolic setting of a beloved site.

As my friend David James Duncan has written, "When your heart's home is being annihilated, your peace and serenity are in deep shit." So what do you do? You can either roll over, or resist. The people cutting the monarch forests are poor, trying to get less so. The man planning to change the valley, by comparison, is wealthy, trying to get more so. The two cases are deeply complex and very different: one concerns poverty and extinction; the other, cultural erosion and economic ambition. All they have in common is the finality of the loss they threaten to exact. In both cases, all we can do is fight like hell, until we win or lose. And then get ready to fight again, because in this world, the heart's home is seldom safe.

Summer 2000

NOTE: We won our valley battle, with many people's help. Our neighbors, the owners of the land in question, are raising fine cattle and hay on the good grass pastures, and have become active in the Grange. The monarchs, on the other hand, are more imperiled than ever in Michoacan by illegal logging, mountaintop warming, and recent freezes and floods; and in the north from GM crops and the poisons they promote. Robust and adaptable but also vulnerable, monarchs and their phenomenal migration will be more and more tested by a world gone wrong.

Alles ist Blatt

Each May I teach in Washington's Methow Valley at the Spring Naturalists' Retreat, an ecological potpourri put on by the North Cascades Institute. While I enjoy the interplay between enthusiastic participants, versatile instructors, and the Okanogan Highlands, the best part is teaching beside Arthur R. Kruckeberg, professor of botany emeritus at the University of Washington. Thirty years ago, Art was our botanical mentor and faculty advisor for our student conservation club. Now, as white-bearded colleagues afield together, Art and I harmonize on the coevolution of plants and insects and joust over their relative importance in the world. Art, or ARK as he is known among his many followers in the Washington Native Plant Society, can always be counted on to issue his favorite botanical aphorism from Goethe: *"Alles ist Blatt"*—literally, all is leaf—or, in other words, everything depends on plants.

A few weeks after the latest retreat, I was conjuring with that assertion on the back porch of a certain Portland brewpub, cask-conditioned IPA in hand. The "porch," a former truck dock, faces a rutted and railed alleyway just north of Old Portland's frontier of gentrification. Having gone straight there from the Oregon Livability Conference, I considered the kinds of features that make this city so famously livable: the pubs, of course; the Willamette River off to the east; vast Forest Park up in the West Hills; Powell's Books a few blocks south. But what really caught my eye were the mellowed bricks and sparrowed corbels of old warehouses not yet metamorphosing into lofts; the elegant old orange rain gutters, set into the brick not as a philistine add-on but as a decorative and functional forethought; the cogs of rusted old pulleys no longer called on to pull their weight. And the green bits, sticking out here and there in the rocky railbed despite the odds.

English ivy, a nuisance at home, softened the far wall, and a tawny patch of foxtail ("tickle grass") adorned a gutter. Against the crumbling base of a building from another era, the intense pink flowers of herb Robert wobbled in the breeze on purple stems. And best of all, decking an iron bridgework above the tracks, hung a wild, exuberant garland of Boston ivy. For years I've visited this granddaddy vine, its six-inch muscular trunks embracing the brewhouse brick. In fall its long danglers make a scarlet chain. Now a chartreuse boa tossing on the river's breath, the *Parthenocissus* reminded me of the famous Hall of Mosses in Olympic National Park. The temperate rainforest's riot of ferns, wintergreens, and oxalis holds little in common with a dusty railroad alley, aside from herb Robert, which grows as an energetic weed along forest and sidewalk edges alike. Yet, what catches my eye in both places is the green stuff—whether the lush giant ears of skunk cabbage, the weird leather ferns on ancient Sitka spruces of the coastal woods, or the formidable thistle emerging from the derelict cityscape.

On a recent hike to the Olympic coast, the most striking plant I saw was bog laurel *(Kalmia microphylla)*, blooming like a throw of deep pink chenille across an olive sphagnum bog. Looking closely at its cerise buds and rose-pink flowers, I saw that the stamens are imbedded in the fused petals, giving *Kalmia's* flower a curious pleated or puckered appearance. Eventually the anthers pull away and rise up to proffer their pollen to the small bumblebees that work them, leaving the pentangular flowers creased by the stamens' former beds.

Just a few days later I found myself in Pennsylvania, criss-crossing the Appalachian Trail near Hawk Mountain. Every time I climbed above the lowland mix of multiflora roses and Japanese honeysuckle that lade the air sweetly (if they do nothing for diversity), I came into a zone of aspen, birch, and mountain laurel *(Kalmia latifolia)*. There was that same distinctively pleated and anther-speckled form, in popsicle-pink buds and apple-blossom-white flowers. Growing on upland soils, these laurels were taller and paler than those of Washington's acid bogs. Yet had I known nothing else about these two shrubs, I could have told from their forms that they shared a common evolutionary and ecological history, until diverging in some far-off icy event. That recognition linked two forests, two sets of plants and their pollinators, across an entire continent.

ARK is right: every place is a product of its plants, as are its people. Our bodies—like those of all animals—perish in the absence of plants' exhalations or their tissues, our minds revel in the intrigue of their lives and forms, and our spirits need their good green ministry wherever we go. At the end of the day, even a zoologist must admit: plants are the bees' knees.

Autumn 2000

NOTE: *Now into his nineties, Art Kruckeberg is still going strong among the plants he loves. Days ago (in May 2012) he called our mutual friend, botanist Cathy Maxwell, to report: "The* Pleuropogon *are in flower here at the nursery!" Cathy had collected seed of the semaphore grass for him for many years, and this was the first time it had flowered for him.*

Carnal Knowledge

I had real compunction about dispatching them in the depth of their passion. They were beautiful utterly merged—and they were stunning in their sheer physical exuberance. But letting them live would have serious consequences, as I had learned bitterly once before.

It was some fifteen years before that I came upon a courting pair of the great gray beasts, the first I'd ever seen. I recognized them from their handsomely spotted hides, and remembered what I had read about their astonishing sexual behavior. So I placed the creatures in a terrarium to watch them. But they escaped post-coitus, and our home precincts have been populated with their voracious offspring ever since.

I am speaking of *Limax maximus,* arguably the handsomest of the large, shell-less land mollusks. Sleekly proportioned, kitten-gray, symmetrically mottled with dramatic black spots, and enormous—sometimes exceeding six inches—the leopard slug is an animal whose beauty must appeal to even the most fastidious slime o phobe. If only it were native! But, like the black and rusty *Arion ater,* which gobbles the garden in its teeming thousands, *L. maximus* originated in Europe. It is one of hundreds of alien species that, having co-evolved with humans over many thousands of years in the Old World, proved pre-adapted for disturbed habitats in North America. So this striking animal is despised as the starling of the slugs, no more loved than the exotic zebra mussels that clog our waterways.

Slugs have a hard enough time with their public relations, but when they are as devastating on the garden as this large herbivore, they haven't a chance at mercy, even from a sympathetic naturalist. Even still, *Limax maximus* practices one of the most dramatic sexual unions I know, thereby seducing our reluctant attention. Like other slugs, the species is hermaphroditic. Every adult both receives and donates spermatazoa, a lifestyle that might be considered highly progressive. But they do not gather in knots of two or three individuals to fuse gonads among the leaf litter as does *A. ater.*

Nor do they engage in mutual penetration with mammoth penises as our indigenous banana slugs do. Instead, great gray garden slugs have contrived a copulatory routine so Byzantine as to raise the most jaded eyebrows.

And raise is the right verb, for these slugs begin their union by climbing a tree trunk or a wall to a high point, then circling for an hour or more, mutually caressing with their tentacles, nipping, and secreting copious gummy mucous. Then, gluing a sticky launching pad to the surface, they drop into the abyss on a shared bungee cord of congealed slime. (One of the greatest attributes of mollusk mucous is that it can be slick as greased glass one second, sticky as super glue the next—an engineering feat no laboratory has successfully duplicated.) There the lovers dangle, like two climbers moved to merge in mid-belay.

Such was what we beheld on a recent midnight, upon going out to the back porch to feed the cats. Ever since that first experiment in voyeurism and the subjects' consequent escape, *L. maximus* has frequented our porch and the adjacent gardens. To their detriment and the cats' patent disgust, they come to the cat food more faithfully than to any slugbait but beer. However, these two fine leopards were sliding up the wall of the house above the catfood dish, more intent on sex than kibbles. We decided to let them reach the ceiling and bungee away, certain we could contain them after witnessing the act. They made remarkable progress in their eagerness. When we checked a few minutes later, they were already slung and linked, their cables pasted to the clapboard wall, their embrace suspended just above the dish. And there stood Firkin the cat, peacefully munching, completely oblivious to the sex play unfurling inches above her head.

The strand hung some three feet, roping the lovers upside down. They wrapped around each other in a double helix so intertwined that we stiff bony vertebrates could only regard their full-body wrap with envy and awe. They dangled and spun, first this way, then that, as their soft exertions spiraled their gyre. And all the while their extruded milky penis sacs—half their total length—pushed out behind their heads, mingling in a clot of blending zygotes. First palm-like, then feathery, these creneled genitals pulsated and throbbed like sea jellies swimming together: stroking, dancing, fanning, swelling—finally forming a sheltering parasol as climax overcame them.

After an hour or so, we humans were exhausted. I'd have left matters there, but I remembered last time, and the ensuing years of heavy plant predation. So I took the copulating leopards by their magical harness, laid them gently in a bread bag, and placed them in the freezer alongside a few dozen *Arion ater.* Later they will enrich the compost, and their magnificence return to the garden of which it is made. Though they would have frozen naturally in a few weeks, making slugsicles now gave me no pleasure and some sharp misgivings. Yet allowing aggressive alien species to reproduce means giving up on native species, as well as the garden. So we make our choices: *Limax* or lettuce, leopards or bananas. I console myself with the thought that there are worse ways to go than entering the Big Sleep in a state of utter rapture.

Winter 2000

Spark-infested Waters

A few weeks before my sixteenth birthday I flew from Denver to New York in an old Constellation airliner with four props and three fins on the tail. I was part of a passel of would-be scientists who would spend their summer at the Jackson Memorial Laboratory on the coast of Maine, courtesy of the National Science Foundation. Our parents arranged for me and Sue, the only other kid from the West, to meet in New York City; then take a bus together up north. She was a pretty and brainy girl from southern California, and by the time we got to Bar Harbor in a cold rain, we were holding hands, and I was a goner.

On a thoroughly magical night during that summer of 1963, our group gathered beside a seashore campfire in Acadia National Park to roast weenies, sing "Michael Row Your Boat Ashore," and watch the Perseid meteor showers. Most of us had long since paired off, and my bus partner and I were in the full throes of young love with the Perseids bursting all around us and the fine calcium-carbonate sand of Shell Beach sticking to our skin.

Then someone yelled, "Look!" and pointed out into the bay, and we all gawked. As if the meteors had hit the water and exploded, the Atlantic was alive with green fire. We had no idea what we were seeing until one of our leaders told us the sea really was alive. "It's plankton," he said. "Makes its own light. Shines just like fireflies."

Fireflies! They did shine with much the same cool green glimmer. There are no fireflies on the West Coast, and Sue and I were rapt with them— just as we were with the smell of the tidal wrack on the seaside rocks, one another, the shooting stars, and now another thrill, the shimmering plankton. One is easily rapt at sixteen. We wanted to run out into the neon sea in the worst way, to be among the green galaxy in the ocean while the meteors arced down overhead. But the Atlantic at Bar Harbor turns your toes blue—especially at night. It would have taken more than a campfire and a beach blanket to warm us up afterward.

Such saltwater fireworks are the bioluminescence of certain dinoflagellates, unicellular marine protists that make up nine-tenths of the rich soup of pelagic life known as plankton. They possess scintillons—packets of luciferin and luciferase, the same substrate-and-enzyme matchbox that oxidizes to produce bioluminescence in fireflies, glowworms, and certain fungi. The best-known genus, *Noctiluca* ("night light"), spirals through the water powered by twin flagellae, and may be as big as two millimeters across. They can make the sea reddish or yellowish by day (hence, "red tide"), but by night, their cold power illuminates the waters blue-green, wherever they bloom in the world's seas.

Over the decades, I have seen the green blaze on the waves here and there around the world, from the North Atlantic to the South Pacific, the China Sea to the Bay of Biscay. And I have met the phenomenon in literature from *Moby Dick* and *Kon-Tiki* to *The Voyage of the Beagle*. But it wasn't until last fall, when Orion's Forgotten Language Tour visited southwest Florida, that I finally had a chance to swim with the sparks.

After giving evening readings, we writers made for the beach behind our hotel. Walking over the famously shelly sand of Sanibel Island and into the bathwater Gulf, we all exclaimed "Wow" in various writerly inflections to see the chartreuse brilliants rise from our steps and circle our legs as if we were human sparklers. I stood still, transfixed, my legs illuminated by key-lime footlights. And when I swam, the amniotic ocean parted in curtains of watery gauze bejeweled with electric emeralds. Diving, there was plenty of light for plucking slippers, whelks, and calico scallops from the seafloor. We ended up splashing one another wildly, giggling as skeins of greeny sequins glittered in our hair and dripped down our limbs, lighting up and flicking out almost faster than we could register their half-lives.

I swam with the night lights every night on Sanibel, barely able to tear myself away for bed. Usually a sinker in water, I drifted easily in the buoyant brine of the Gulf, suspended by water and light—or as one Caribbean captain put it, "floating in stardust." And when I clapped my hands overhead, showers of shooting stars fell all around, transporting me back to 1963, another bay, and a night alight with fascination, infatuation, and a summer storm of living sparks.

Spring 2001

Durable Goods

In this coldly mercantile era, economists clutch at any straw in the wind to woo stability from chaos. Among the "leading economic indicators," the one I like best is "sales of durable goods." Whether or not this measure helps the Fed set interest rates, it certainly furnishes a fine example of unintended irony.

When I think "durable," I picture a friend's '47 flatbed Ford, kept running with TLC. Instead of an old truck, I drive an '82 Honda Civic hatchback named Powdermilk. We've gone farther than the moon together: 289,000 miles and counting. Recently, unable to ignore the blue smoke we were contributing to the Lower Columbia airshed, I had her engine rebuilt. Dave did the deed at the shop where I bought her new and have always gone for service. Some saw this act as more sentimental than sensible.

I plead guilty to a loving attachment to every automobile I've ever owned. First there was a hundred-dollar '50 Ford with overdrive. How my friends and I survived shooting over prairie and plummeting downcanyon in this seatbeltless, primer-gray pellet, I'll never know. That venerable vehicle preceded a dynasty of Volkswagens—two Bugs, two Buses, and a Squareback—before my fidelity shifted to Japan. I recall the carriages themselves, their charms and peccadilloes, but also the field trips we took: to the 14,000-foot top of Mt. Evans in the old Ford; 10,000 miles birding across Europe in a red Bug; another summer's trans-American trek in an ancient gray one before it succumbed to Washington mud; camping in many a mountain range and desert canyon in both rusty buses. I remember the birds I hit with each car, and the rainy trailheads where my faithful wagons waited, patient and dry.

But it's not sentiment alone that drives my automotive allegiance, any more than affection made me plunk a sluggish two-dollar modem into my computer to keep it connected, rather than upgrade as everyone said I should. Nor is it entirely a matter of expense; I spend far more than my mechanical brother would to keep a car on the road. No, my contrarianism is deeply ingrained: when the chorus calls, "upgrade," I shout, "up yours!"

Bigger and faster machines are even better at cutting our links to the natural world. Fortresses on wheels ensure against any real contact with the countryside, and it's hard to see much with the cruise control set at 70. But slow it down to zero mph, step outside the aluminum womb, and everything comes to life. One stuck-in-the-sand episode gave my best look ever at the elegant, brassy Behr's hairstreak butterfly, when a mating pair flew in the window and alighted on the hot, dusty dash. Likewise, ingenious Web sites beguile the hours, and absent restraint, whole lives. Yet when the screen goes down, actual experience blasts off. Days unplugged are days connected to reality.

Now, I am not a very good Luddite. Many live more lightly than I— off the grid, or with bicycles for wheels. I know several well-published writers who abjure computers altogether. I too prefer writing manually, but after composing a 924-page field guide manuscript on a typewriter and nearly perishing of white-out poisoning, I embraced a motherboard. And since relying on public transportation in my rural district means staying home, I keep an auto. But I have not evolved as rapidly as automotives and electronics. I don't have a TV, microwave, answering machine, cell phone, or laptop (except an old leather timber-cruiser's bag that holds my journal and does, actually, work just fine on my lap—in woods, pub, or airport). Not that I feel smug, living without these devices; sometimes I miss them. It's just that in a day and age when some insist we are already cyborgs— creatures of our own machines—I would rather be a creature of creatures.

The idea that we must jettison large bundles of metals and plastics just to buy fancier ones is deeply repugnant to me. My Neolithic computer (last week it was merely Bronze Age) renders acceptable manuscripts in runes of both ink and electrons; then I turn it off and go outside. My rebuilt, gas-sipping buggy gets me where I'm going, be it 3 miles to the post office or 9,000 miles after migrating monarchs, until I cut the motor and go on foot.

Why shouldn't a '47 Ford truck, made of steel and hard wood, last as long as a '47 Human like me, fashioned of soft flesh and bone? The giant vehicles spilling out of lots and onto highways today fire up the economy, but fit poorly into a world with limits. Ecologists know that frugality begets stability, while animals that consume extravagantly, like cinnabar moths on tansy ragwort, boom for a while and then inexorably bust. Different strokes for different species, but an animal with a choice might wish to avoid the discomfort and inconvenience of periodic extinction.

An economy based on perpetual upgrade and disposal cannot possibly persist in a finite world. As I see it, a culture with a chance at a long and healthy life is one that speaks of "durable goods," and means it.

Summer 2001

NOTE: *Powdermilk the Honda, now thirty, drives on! As of summer 2012 she has 408,000 miles on her odometer, and still gets over 40 mpg.*

Lookee Here!

My grandfather, Robert Campbell Pyle, was famous for finding money. Each year, the *Rocky Mountain News* reported how much Mr. Pyle had found since the previous dispatch, usually more than $100. Then he would break the take out of his plastic baseball banks and old socks, depositing equal shares in his grandkids' savings accounts. GrandPop didn't go in for a metal detector, nor did he turn up any Spanish gold coins or Confederate scrip from old attics. For him, it was nickels and dimes, quarters and pennies, and very rarely a half or silver dollar, plucked from the streets of Denver.

I remember GrandPop in white shirt, straw hat, and watch chain, and usually a coat and tie except on very hot days, looking down, stooping over, and always coming up with a coin. Strolling in City Park or idling between his boarding house and favorite cafes on Colfax Avenue in East Denver, he was ever watchful for loose change stuck in the hot asphalt or lurking in a sidewalk crack with the cheeseweed and gum wrappers. Every find was met by the same exclamation, along with a slight chuckle of satisfaction: "Well, lookee here!" I thought that was just his saying, until much later when I recognized "lookee" as a contraction of "Look ye," the old formal English having hung on in GrandPop's Kentucky long after it dissipated elsewhere. I was awed by his money finding, and believed it a matter of luck. Now that I find myself copying his habit, I realize that my grandfather's pick ups, as he called them, owed mostly to the simple act of paying attention.

Vladimir Nabokov said, "It is astounding how little the ordinary person notices butterflies." For animals equipped with remarkable sensory tools—less acute of eye than falcons, of ear than cats, of nose than dogs, but still darn sharp—we are an oblivious lot. Just as I missed the tarnished dimes and dull nickels that caught GrandPop's eye, and others often overlook the butterflies I spy as a matter of course, many people are blindered to much of what goes on around them. Nabokov's plaint might well be rephrased to say, "It is astounding how little the ordinary person notices *anything.*"

You may simply say "search image" and dismiss it at that, and to some degree you'd be right: by honing our ability to pick out certain forms, we succeed in spotting the desired objects more often—as people do with birds, or scarce Model A parts at flea markets. It is also true that some people are naturally more observant than others. Take the journal artist Hannah Hinchman. When groups hike with her, they are routinely abashed and delighted by the elements of the landscape that her eyes tease from the whole. Of course, this trait improves both her art and her effectiveness as a naturalist. Such perceptivity serves the writer's task as well. No one's poems, essays, stories, and novels are more richly dressed in nuance than those of Nabokov, who exalted the "individuating detail" in service to his art: as Humbert and Lolita passed through all those motorcourts, the American landscape rolled by in the first-person particular, trained under the same microscope that the author brought to bear upon butterfly scales.

But most of us, by editing out the clutter of modern culture, miss the good stuff too. When Starbucks logos eclipse backyard beetles, birds of passage, even the stars themselves, the natural world becomes a quaint abstraction, of little familiarity or value. "Unreal," you might say—as one gentleman recently described a spectacular canyon we were both regarding. And when people do take note of something common but wonderful, they often conclude that it must be "unique," or "extraordinary," or even "a mutant." Every spring someone brings me a "new" butterfly, which invariably turns out to be the handsome but abundant inchworm moth, *Mesoleuca gratulata*. I too am stunned now and then by some natural feature that had to be there all along—a beetle, a lichen—unable to accept that I have missed it until now.

Not many of us will ever acquire GrandPop's preternatural eye for small metallic disconformities on the city streets. But just as the scales fell away from Saul's eyes, restoring his vision, we too could rediscover the gift of sight. This matter is of no little consequence. What we truly see, we have a chance of loving; and what we love, we might hold dear.

No, I'll never be the money finder GrandPop was; usually I pull in five or ten bucks a year, unless I make a lucky connection with a stray greenback or two. But peering closely at the ground as well as the sky, watching for

Lookee Here!

shiny Ikes, Abes, Toms, and Georges, I also see the flowers that flout the concrete's ban and the cities of ants below our own. I want to make things real. I want to live in a world that's not beneath our notice. I want to say, "Lookee here!" every time I open my eyes.

Autumn 2001

The Toucans of Tikal

My friend Jimbo traveled to the wild Petén district of Guatemala to study bird behavior in the early '70s, and his description of bat falcons shooting around the ancient Mayan ruins at Tikal has stuck with me ever since. So when Thea and I took a keen troupe of butterfly watchers to Central America recently, I climbed the temples at Tikal and watched for the falcons. Swallows gave false alarms, appearing and disappearing around the gray pyramids, and black vultures perched on top like gargoyles, waiting for us to fall.

From the high terrace of Temple Four, I listened to packs of howler monkeys roaring unbelievably from two directions, as loud as prides of lions in wraparound stereo. Later, in the rainforest below, a dozen coatis—long-snouted, long-tailed relatives of raccoons—rooted at my feet. Overhead, keel-billed toucans presided.

The trails were almost deserted, air travel having shrunk to a trickle in the aftermath of the New York enormity. We too had considered canceling our expedition, but carrying on with this peaceful and reverent journey seemed a respectful response to tragedy. Seeking solace and serenity in soft bright wings, we had the normally crowded shrine almost to ourselves. So I watched the toucans undisturbed, and found myself asking how this big, rainbow-billed bird could be taken seriously, after all those Guinness and Fruit Loops ads. How can any animal or plant retain its dignity, when its image has been co-opted by commerce, stylized, and universally broadcast on billboards, cartoons, and television? This is a serious matter, because proper citizenship requires us to regard other life forms with care and compassion. It is also a selfish concern: when I meet a new species, I want the experience to be fresh and passionate, not depleted by commercial baggage. In short, I wondered, can a creature become a cliché?

These questions are particularly pertinent to that famous trio of southwestern pop mythos—roadrunner, coyote, and saguaro—mostly due to those Looney Tunes, with Wile E. Coyote chasing the Roadrunner past the same three or four hands-up saguaros, endlessly. Coyote is also blazoned in the brain as pastel song-dogs marketed between priapic Kokopellis and Chinese-made Kachina dolls. And saguaros! How many greeting card cacti must we glimpse before a live one ever graces our vision? I worried that I would see saguaro as little more than a bristly silhouette from a comic strip; that a child raised on cartoons and commercials might miss the actual encounter, waiting for the wheel-spinning bird to emit its trademark "Meep, meep!" as a skinny, bipedal, snaggle-toothed coyote hugged its tail with mayhem in mind.

But it hasn't worked out that way for me. Last summer, returning from a rangers' reunion in Sequoia National Park, I spotted two roadrunners crossing—what else?—the road, then pacing alongside a citrus grove and disappearing within. The big speckled bird, with its wild yellow eye framed by blue and red, its scimitar bill and erectile crest, conveyed the thrilling sense of an organism's perfect fitness to its place. I thought of the blue, snub-nosed cartoon version no more than I hear Woody Woodpecker's shrill guffaw in the cry and hammer of a pileated. And when I finally saw saguaros in 1992, thrusting up through yellow composites, pink penstemons, and mimosas, poking the hot air, they failed to evoke the inflatable cacti that live with pink flamingos in novelty stores. And as for coyote, the one that limped across the highway near home, hit, holding its shattered paw like some vole it was taking to its kits, was no Wiley. The ones that tremolo and countertenor in the plum grove on September nights are no cutout songdogs trivialized by some southwestern chamber of commerce.

So what of those toucans? The magnificent green, red, orange, and blue schnozz briefly brought to mind the old Guinness ad, me being more inclined toward Irish stout than Fruit Loops. But when the big black and yellow bird stretched its massive bill down to preen its belly, so the scarlets of the tip and the undertail coverts met, both backlit by the sun, I knew that no caricature could steal an organism's singular evolutionary elegance. If anything is to be co-opted, it is our own perception. This we must resist, for when animals and plants are trivialized, so too are the ecosystems they inhabit.

On the other hand, maybe the pop-copying of critters can have some redeeming value: millions of people who don't know a sparrow from a swallow might recognize a roadrunner if they saw one. As long as cultural exploitation of living things does not debase or demonize them, perhaps it actually helps maintain connections with other species in a world where we sever them at every opportunity.

I never did see Jimbo's little orange falcons knifing over the high stone temples in the jungle. But in their stead, the toucans of Tikal taught me this: banality is a bane of our own creation. It is up to each of us to keep the world fresh.

Winter 2001/02

Waving the Flag

I've been thinking a lot about flags lately. It's been hard, after all, to do otherwise. For example there is the blue flag iris. The name usually refers to *Iris versicolor* of the North and *I. virginica* of the East, but I am better acquainted with those wild marshy banners of the old frontier, *Iris missouriensis,* often called western flag. The sight of a single Missouri iris waving over a wet Idaho meadow last summer put me in mind of an afternoon many years before in the Colorado high country above Nederland. Brillo clouds had come up, closing out the possibility of butterflies. Then, rounding a curve on Highway 7, I beheld a broad lap brimming over with blue.

Until then, I had mostly known the fancy irises of my mother's and grandmother's gardens, numbering in the dozens. But here I beheld irises by the thousands, just as handsome in their wild simplicity. I pulled over to walk out among them. Each narrow, pale blossom was occupied by metallic blister beetles—shimmering deep blue-green to bronze to purple—jumping each other's exoskeletons. I'd never seen such a randy, brilliant bunch of beetles in full throat (if only you could hear them!), nor had I beheld such a sweeping field of flags.

Then there is the posterior pennant of the white-tailed deer. While visiting Virginia this past fall, I hiked with friends through a tunnel of red maples up to a high stone bald. The white granite of the Blue Ridge held clefts of fern and lichen, clumps of russet scrub oak, vistas of ranked hills in autumn motley, rolling off into every distance. Hawks and monarchs were on the move. Once we left the summit, all the colors bled to browns, yellows, oranges, reds, and lingering greens. I let the talkers go first, so as to take in nothing but the leaves falling, leaves underfoot, their swishes, snaps, and smells. Then, *Wham!* A cinnamon flicker of fur erupted from a bittersweet brake, chased by a white-hot candle. Once it was raised, I could no more take my eyes off that bobbing, flouncing flag than follow it, as it blazed a way through the autumn wood.

A white flag often means "truce" in human affairs, just as a red flag signals danger ahead. But for other animals that can discern it, red holds subtler shades of meanings: the bitter insect's "don't eat me," the succulent berry's come-on. When birds flash red, it can spell attraction for mates or repulsion for rival males. Among my favorite flags, the scarlet epaulettes of the red-winged blackbird probably do both. And the "Ocaree!" of the male red-wing calling from atop a cattail never fails to make me stand up straight and salute that flash: a whole cardinal's red concentrated in one patch of flaming feathers.

But nature's presentation of the colors is hardly limited to reds, whites, and blues. Especially in the monochromatic months, our eyes hunger for hue, and the wait can stretch out. Something tells me that this winter is going to be a long and a cold one. There will be floods, and damage done, and losses; gray will stay till the cows come home, and there are no more cows. When the fog's dirty cotton dressing finally falls away, we will all be desperate for vital signs. So when they finally appear in the sodden pastures, reflecting the return of the sun from the southern sectors, nothing will be more welcome than the upstart flags of skunk cabbage. Unlike the greeny-purple twists of northeastern skunk cabbage, the western species unfurls its spathes into broad, tall swatches of yellow—the uncompromising yellow of early dandelions, pioneer daffodils, even buttercups. No emblem commands my allegiance more deeply than these, announcing the cycle's rounding once again.

Finally, ushering in solid spring with their full-staff flurry of green, come the ensigns that uphold the security of all life: the leaves. When the soft fresh spears of Indian plum first unwrap their packets of incredible freshness, followed by elder, alder, maple, and ash, the entire citizenry of the land awakens unto erectitude, able to face another season after all. There is no older glory than this. April to August, the leaves in their flapping, waving, snapping, and growing flaggage raise deepening green over the countryside, week by grateful week. Then, by measure, they grow tired, dusty, and mildewed, scalloped and mined by insects, summer blasted; and then ... fall.

But they will be back. As will the skunk cabbages, red-wings, white-tails, and irises. Through the rocket's red glare, these standards never flag. Long may they wave.

Spring 2002

My Meringue Bazooka

Deep in the Dark Divide of the Gifford Pinchot National Forest, a bunch of botanizers from the Washington Native Plant Society is exploring the high meadows in midsummer. We've backpacked up to the saddle between Sunrise and Jumbo peaks for two days of flower forays and the kind of campfire chats that naturalists love. I am losing myself in alpine air and heather scent when the montane silence cleaves before a wave of sound that begins as a hum, grows into a scream, and then an intolerable wail. It is a hive of motorcycles—not the Vinson '52s you might see cruising Highway 66 or the Harley Hogs on US 40 bound for Sturgis, but high-assed Hondas and Yamahas—tearing Boundary Trail #1 to bits.

This sevensome of dirtbikes roars right up to the edge of a trail-blocking snow-field, revs around, and rides smack through the delicate heatherbeds, ripping out shooting stars and sedums in the rototill of their knobby caveman wheels. As they rejoin the trail and roar away, their racket and scramble split the vista, scatter the mountain goats, and tear up the turf for as good as all time. The wild alpine air closes in again around the damaged day, dispelling their gasoline fumes, if not my anger.

In his epic book *Voyage of a Summer Sun,* Robin Cody recounted a similar experience at the beginning of a canoe trip along the entirety of the Columbia River. High up among the wild headwaters in British Columbia, he wrote: "The whine of jet skis shattered the thin air. Two grown men, their wet suits filled to capacity, throttled past and unzipped the river. ... Waiting for birdcall to come back to the river, I reflected on the whole idea of jet skiers and why we should let them live."

I confess to similar sentiments. But as an unreconstructed adherent of nonviolence, I have searched and searched for ways to deal peacefully with jet skis shattering wild lakes and rivers, snowmobiles trespassing on the white silence, and leaf blowers anywhere at all. I have languished on many a college campus, where of all places one should be able to enjoy languorous afternoons with a book and a breeze, only to have the serenity shredded by these satanic devices. No culture in which leaf blowers prosper can be anything but in decline.

In contrast, I recall a monastery in Moscow where the dominant sound was the gentle whisk of birch-twig besoms on old flagstone walks, as an inbred strain of brown cats sparred with hooded crows in an ancient Slavic dance, vying for bits of bread left behind by babushkas in black skirts and scarves. And just last fall I heard the rare, romantic music of the rake, as groundsmen in Guatemala cleaned leaves and monkey droppings from the lawns around Mayan temples.

The problem, whether in the backwoods or the heart of civilization, is the theft of the golden silence by selfish or expedient din. On a Forgotten Language Tour in Ashland, Oregon, I was almost drawn into a rumble by a belligerent with a leaf blower across from my hotel at 7 a.m. That's when I decided to put my plan for nonviolent resistance into action. Now I am quite happy to let the boors live, thanks to my meringue bazooka.

This blunderbus projects a gobbet of ultra-compressed meringue from its barrel. The wad homes in on sharp sound waves, so the shooter can hardly miss the malefactor. Upon contact, the dollop expands into an enfolding carapace reminiscent of those swirly wedding reception cookies that no one seems to eat, but much bigger. The hardened meringue fully envelops its target, but remains porous, so offenders can still breathe. Air-filled, it floats like kapok, so no one drowns. Targets can, of course, eat it too, so starvation is not a problem. They just roll about in the sweet embrace of the meringue, absolutely silent. The MB also promises to ensure better fare at wedding receptions by using up all the available meringue.

I heard on the news recently that a retired professor in Wales has invented a "sound shadow" that blots out unwanted noise by generating the opposite waves. Apparently it works, but will be expensive, and useless for moving targets, unlike the MB. However, it shows I'm not alone in my disdain: sound pollution must be among the most irritating unsolved evils of the industrial age. Then there is the severe fouling of air and water accomplished by two-stroke engines. You'd think everyone would want to fight it, but the noise just gathers, louder and louder. Luckily, I've found something even better than the mute button on the remote control.

I can see the MB's extension into realms of warfare and other asymmetries of human behavior. Instead of casualties, we will collect meringued hostiles bouncing around the theater of conflict, awaiting peaceful repatriation. This device might even revolutionize revolution, as one could presumably use it on dolts, scoundrels, and ne'er-do-wells in various governmental bodies, cartels, and gangs.

I've got just a few bugs to work out in the meringue bazooka before production begins. How, for example, to keep it out of the hands of my neighbors, when I fire up my 24-inch Stihl.

Summer 2002

Nightlife in Amsterdam

It was the month of March. Logan, Utah, where I was teaching, still lay in the chill grip of snow and ice. A trip to Holland for a butterfly conference in the ancient university city of Leiden seemed the perfect leap forward into spring. Before departing, I asked my environmental writing students to read an essay of Brian Doyle's entitled "Van," to study its near-perfect form, and to ponder the way Van Morrison's music ties together the human heart and the natural world. Graced with images from the rivers Lagan and Beechie and the woods around Belfast, bristling with crickets and bogs and boyhood haunts, his songs show how waters and rocks and love and desire all spring from the same source. Singing "Take me back, take me way back ... when you walked in a green field ... through the buttercups, in the summertime ... When I understood the light," Van testifies that nature and culture are two sides of the same coin. But I wondered if this would still seem true in the densely populated Dutch landscape.

After the butterfly meeting, I bicycled along canals through the heart of the famous flower-bulb fields around Leiden. The early spring banks were alive with golden celandine, and the still water exploded in the courtship displays of great crested grebes. Picture long-necked, spike-billed dragonets in chorus-girl headdresses with fancy ruffs of amber and black, foxtrotting *a deux* across the water. As the great tulip-red sun settled behind a village church at the end of a long yellow band of daffodils, my sense of being in a mature countryside—a landscape concocted of equal parts nature and culture—was profound. I was well aware that centuries of filling wetlands to make these Dutch polders had eradicated many lifeforms. Yet the present condition supports plants and animals in broad array amid an attractive and prosperous human community.

This perception of ancient settlement, of an intimately shared (if not always peaceful) history between human and other inhabitants, only increased when I visited the Belgian city of Bruges. Following the barging moat on foot, past half the medieval city gates and all four windmills, then out beyond the old row houses with their fancy gables, I came to a woodside field where town and country merged. Rabbits boxed and lapwings soared and dive-bombed, all out of sheer sex. The lapwings, wispy-crested, round-winged plovers, shone with all the colors of violet-green swallows. They shrilled their signature "Peee-Wit!" as they rose and swooped, over and over. The first butterflies—lemon-wedge brimstones, a puffball cloud of speckled woods—were also seeking mates between nectaring bouts on sprays of coltsfoot. Walking back into town at dusk, I paused often to watch swarm flies swirling like atomic particles, the males presenting prey wrapped in bright bubbles as nuptial enticements for the popular females.

The next day, as the train carried me east through Ghent and Brussels, then north past Antwerp, Delft, Den Haag, and finally into inevitable Amsterdam, the gathering urban/industrial clot strained my sense of harmony. The Dutch capital would be nobody's first choice for a natural history destination, though it does enjoy an abundance of trees, parks, and waterways. But I'd purposefully spent my spare days in the country, arriving in the city just before dusk with plans to fly out the next morning.

The Concertgebouw Orchestra was sold out, but imagine my surprise when I discovered the other main event in town that night was Van Morrison! I made my way by metro to the Carré, where he was playing, only to find the house full. Explaining my situation to the usher in the small circus hall, I asked if I could just listen for a moment from the lobby so I could tell my students that I had heard Van. She did me one better and let me stand in the back of the hall, "for just five minutes." After a couple of numbers, another usher came to show me out. Just then a striking blonde saxophonist took a solo, and the usher became transfixed. "She is Holland's most famous musician right now," he said, clearly her most devoted fan.

"Would you like to use my binoculars?" I asked, handing him my trusty pair. I always keep my binos at hand. I've found them equally useful for whatever comes my way, be it great crested grebes or the Grateful Dead. Of course I saw the whole show after that.

I'd never seen Van Morrison live before, but I had listened to his remarkable voice with care and devotion since high school, and I was not at all surprised that he was wise and thrilling and cut to the heart.

Afterward, "Carrickfergus," a haunting traditional song he once recorded with the Chieftains, echoed in my head as I floated along the banks of the big city canals. High on my free concert and buoyed by a good Belgian beer, I seldom touched solid earth. But how to spend the rest of the night? I wanted to walk to Schipol airport, birding as the sun rose; but it was just too far, given my early flight. I turned toward the harbor, where all rides to Schipol originate.

Amsterdam is like half of a bicycle wheel with ribbons woven through the spokes, where both the spokes and the ribbons are canals. As they near the hub (the harbor), the spokes converge, funneling walkers through a crowded, boisterous quarter that turns out to be the city's noted fleshpot. From the broad, airy canals and boulevards, narrow buildings lean over narrower alleys, and the ceiling closes in beneath a tobacco-and-cannabis haze. Bands of rowdy young men and older singletons cruise and troll the sidewalks, eyes glazed with silicon visions. But there are also well-dressed couples, even families, and people dining in nice restaurants wedged into the warren of bordellos. And throughout it all, swim-suited women pose behind windows framed in red neon.

I tried to compare the transacting parties here with the courting creatures of the countryside. Were the strutting, gawking, and posing not tantamount to the garish dance of the great crested grebes in the canals? Didn't the big hair and silicon blandishments of the women hark back to the headdresses and puffed breasts of those outrageous waterbirds? Don't the whipping topknots and iridescent plumage of the lapwings remind you of the razor-cut 'dos and expensive clothes of the eager customers? And the shiny euros changing hands bring to mind the bright bubbles that male swarm flies present to their potential mates.

It's all biology. But despite their entertainment value, the comparisons left me unconvinced. That guide books treat the whole scene as just another tourist attraction—"See the Rijksmuseum, see the windmills, see the pretty ladies in their cages!"—stuck in my craw. Nothing tangible tied this Dutch not-so-still life to the wider sphere beyond our own manufactured world. That centuries-old Low Countries ideal of countryside as cultured landscape and city as peopled habitat was stretched thin indeed.

The March night was hot and close, and I couldn't seem to find a way off that smoky, strident island. Then a tiny bridge appeared over a little canal whose fetid water reflected the pink glow of the neon. And there, among broken boats and tossed-off trash, paddled a coot—a plump gray pellet with an ivory-white, humpy bill and big feet splayed like horse chestnut leaves. Now, I've seen coots all over—I just hadn't expected to run into one at midnight in the red-light district of Amsterdam. Splashing away into the murk, its scuts flashed bright white as it vanished down the garbage-clogged canal between the lurid lanes. It was only a gutter coot. But its brief, extra-real appearance took me back to every coot I'd ever known, in marsh and river and lake. And it somehow linked those women, flashing their own scuts at the randy and curious passersby, with every habitat outside their crimson neon niche.

I took a sleepy bus ride past many other faces of Amsterdam, then a long pre-dawn walk around the country-industrial moat surrounding Schipol. Napping on a wooden bench, I heard Van's voice in my mind—"soft is the grass and my bed is free"—and saw rectangular visions of bright bulb fields and glittering canals. When I awoke, it was to the shrill whistles of jets taking off and shorebirds piping out of long grasses, the last of the night's life, in the wan light of Easter morning.

January/February 2003

NOTE: *The blonde saxophonist is Candy Dulfer; you can easily find her dueting with Van on* You Tube.

Wings to Wander

Three weeks ago I waded in Atlantic waves before a pink Art Deco hotel in St. Pete; the very next day, I bodysurfed in the Pacific off Oahu. After basking on a slope of straw-gold bunchgrass beside a cold Cascadian lake at five thousand feet, I boated the Rio Grande in south Texas, then crossed into the Sierra Madre Oriental, in search of subtropical butterflies and birds. Now, in my mind's eye, blue-green agaves morph into October aspens colored every shade of citrus, as waist-high orchids on the rim of Kilauea Crater grow into tangled Floridian mangroves. White ibises worm a tropic lawn; scarlet 'apapanes nectar crimson blossoms of 'ohi'a trees; mourning-suited nutcrackers probe ponderosa pines; and maroon-fronted parrots garrote the Mexican air against a high limestone wall: Tampa to Tamaulipas, in a flash of bright wings. Now, heading home in the 737 *du jour*, the only wing I see is that broad aluminum plain sweeping over a snowy Rocky Mountain cirque.

If you'd told me as a teen that I'd be living this way, I would have shouted for joy. I always hoped to be a footloose naturalist, wandering hither and thither, chasing fabulous creatures to ground. What could be better? Looking at any map was sweet torture; I longed to be in all places at all times. I still thrill to fresh exploration, though nowadays I also lust after my own literal and literary backyard. Yet as much as I aspire toward the ethic expressed in my colleague Scott Sanders's book, *Staying Put*, it's just not in the cards. I bounce about like a free radical let out for recess, sucking up frequent flier miles like oxygen.

As both writer and scientist, I am doubly tugged away from my moorings. First, since writing books falls far short of a livelihood for most authors, we become migrant workers, dancing across the countryside from gig to gig, lecturing, teaching workshops, keynoting conferences, leading groups afield, giving readings. Bids for these odd jobs usually come from afar, and not always in convenient succession: one itinerary last summer had me dining with Vermont fireflies one week, dancing with Arkansas chiggers the next, then back to a Vermont bog for butterflying amid the blackflies. Second, the creatures I study inevitably draw me to their home ground. Surrogates in books and Web sites just can't sub for firsthand, face-to-face encounters. A Mylar-blue morpho in a butterfly house is an impressive thing, but it's not the same as a wild *Morpho peleides* sailing out over the Costa Rican Caribbean, a dazzling blue-on-blue. The evergreen zing I get from meeting new species in new places still fires my get-up-and-go as sure as anything I know.

Of course, the getting there must be endured. Fortunately, even the airplane cabin offers something for the naturalist, beyond the opportunity to study human nature under boredom or duress. I always scramble for a window seat, taking great pleasure in the landscape as seen from thirty thousand feet. Some people consider air travel too fast and artificially distancing to enjoy, as if only overland travel "counts." But what better way to take in the texture of topography than to soar above it like migrating cranes or monarchs: the watercourses, the dips and folds of the hills, the maroon and lime salt pans of Great Salt Lake, the tangent green disks of the circle-sprinkled crops. That Colorado cirque we've just passed—scooped out as by an ice-cream dipper, now cupping a vivid blue tarn—alpine glaciers once squatted there, growing and carving over ages, their granite talus assuming the angle of repose and making habitat for pikas and Magdalena alpine butterflies.

On many Atlantic crossings, you can still see such cirques being cut in the *T. rex* jawlines of Greenland, sometimes shining lime-green beneath the aurora borealis. Nothing mystifies me more than a cabinful of people locked onto their laptops as Greenland goes by; nothing frosts me as much as being asked to drop my windowblind for the video as the great land passes beneath—unless it's being stuck in an aisle seat when the purblind window-sitter pulls the shade down, movie or not.

To be sure, all this travel has its downsides: the bottomless suitcase, the endless succession of airports with departures (what with security) clustering ever more toward the predawn, the poor forage, and excessive sitting for all those hours in aluminum tubes pressurized with bad air. But frequent flying has greater consequences than pretzels for lunch, dreadful magazines, and the real drag of jet lag. In a profession devoted to reverence for place, is all this shifting about really a good thing to do? I sometimes question my capacity for taking it all in, for seeing what Nabokov called "the individuating detail"—how many trips until it all runs together? Then too, it's devilishly hard to find unbroken writing time when you're always preparing for the next departure. Sometimes the trip I most yearn for is the return. After all, according to Thoreau, the only good travel is that which makes us happy to be home.

More important, all this burning of airplane fuel does large harm to the airshed, the ozone, and the fossil fuel reserves. When the recent building of Denver International Airport erased habitats from my childhood, I was one of the millions of reasons for its construction. Far-flung environmental conferences, where I often speak, strike me as especially ironic: all these well-meaning activists migrating thousands of miles to discuss and consider ways to diminish the ill effects of industrial society, whereby too many people use too many resources. ... Does the good we do justify our globetrotting? We'd like to think so.

Fact is, I am really a train man; I'll take the singing rails over the whine of jet engines any time. A recent book tour of mine was conducted almost entirely by train. As soon as I step aboard I enter a state of utter relaxation. And the land and rivers are right there at eye level: I identified eighty species of birds on one winter's round-trip from Seattle to Denver and back. I knew goldeneyes from Washington, but I had no idea they wintered on the half-icy rivers of the western mountains. From the observation car of the canyon-hugging train, I clearly watched the warm brown ducks and pied drakes with a snowball patch on their coal-black cheeks, bobbing for fish between puffy white banks. But the airline-bailing Congress is intent upon the forced starvation of Amtrak; so mostly, we fly.

As we begin our descent into Washington's verdure, I think of the green kingfishers, the green jays, and the green malachite butterflies like animated alder leaves in an autumn wind, but neon. I try to keep each image separate and complete. This way of life can't go on forever—even pelagic limpet spawn eventually settles down for the long run. I suppose I will too.

Meanwhile, even as I groan over every departure and long for each return, I know I protest too much. We who roam afar should always pack a keen sense of privilege, and have the good grace to wonder at the wide world.

March/April 2003

Confessions of a Hophead

During the Q & A after a recent lecture, a man in the audience asked me whether any butterflies feed on hops as their host plant, like monarchs and milkweed. "As a matter of fact," I replied, "yes!" While I shuffled the facts in my mind for an accurate response, I found myself smiling at the mere mention of that golden word: hops! One of the great ingredients of life, and long recognized as such. It's not known as "the noble hop" for nothing, and that moniker has nothing to do with rabbits or rap. Not even with be-bop, though the first song that ever made me want to get my kicks dancing with the chicks was Danny and the Juniors' "At the Hop."

No, this hop is an aromatic plant. Its scientific name, *Humulus lupulus,* is about as hard to say as "brewery"—especially after visiting one. And that's where you'll find the noble hop by the bushel, for hops are one of the main components of nearly all beer brewed today. In fact, hops are one of only four ingredients in beer that's made traditionally. Germany's sixteenth-century *Reinheitsgebot,* or beer purity law, limits the fixings to pure water, malted barley, yeast, and hops. Nowadays many big brewers substitute cheaper rice for barley, some recipes call for wheat or oatmeal as the fermentable, and boutique brewers emulate the Belgians' penchant for cherries in their beer by adding such impurities as raspberries, honey, pumpkin, and spices—all fine on their own, but nothing to do with beer, in my palate's opinion.

Being a botanical, *Humulus lupulus* originated in the wide world beyond the brewer's yard. It belongs to the marijuana family (Cannabaceae), but its finger-lobed leaves more resemble those of the related mulberry or figs. The common hop abounds in open habitats around Eurasia and is naturalized in parts of North America, where it is sometimes hybridized with the native subspecies *H. l. americanus* for the taste and hardiness it brings to the mix. Linnaeus named the genus *Humulus* from the Latin for hops, the species *lupulus* for its habit in Roman times of growing wild among willows like a wolf among sheep. It is a long-lived perennial vine with chartreuse, papery flowers or "cones" borne on the female plants. These possess glands called lupulin, which produce the complex volatile oils and resins coveted by brewers. The ancient Greeks employed hops medicinally to calm digestion and ward off leprosy. For Native Americans, they served as everything from a pick-me-up to a sleeping pill. Hop shoots have been prepared and eaten like asparagus, and while the hop tea I once tried was fairly wretched, baking brownies with hops might be worth a shot.

Hops, cultivated for beer making at least since the eighth century, serve two purposes in brewing. The first is preservation, since their essential oils have a naturally retardant effect on microbes that might spoil beer. The second is flavor: hops, in a word, are bitter, which is why the basic British ale is known far and wide as "a pint of bitter." Malted barley, when fermented through the action of yeast, turns into soluble sugars and alcohol. In the absence of a bittering agent to provide balance, beer would be cloyingly sweet, sticky, or heavy on the palate. By boiling hops into the wort (an aqueous infusion of malt) "the desired mellow bitterness and delicate hop aroma" are imparted to the beer, according to the 1956 *Britannica*. Before hops, beers were probably molasses-like, or downright dreadful. Coriander, bay, and juniper served as bittering agents at one time or another, and when hops weren't available on the frontier, new-growth spruce tips sometimes sufficed. Spruce beer is still brewed by Siletz Brewery on the Oregon coast, and by moonshining loggers and hoedads practicing the gentle art of zymurgy in the foggy evergreen outback of the beer-rich Northwest.

Hops are an acquired taste, but once acquired, much beloved—a "righteous joy," in the words of the Stone Brewing Company. Botanist and chili-lover Gary Nabhan has told me about over-the-top chili addicts who pop the supra-hot wild chilipequines as if they were popcorn. Similarly, your real hophead nibbles his herb raw, and can be seen snitching hop flowers on brewery tours. The humulophile's preferred nectar is a subspecies of English bitter known as India pale ale. During the days of the Raj, British authorities had to ship ale around the cape to the subcontinent for Her Majesty's troops, and it often went bad in the tropic latitudes. Brewers found they could prolong its life by making strong beer in the city of

Burton-on-Trent, whose waters contained a lot of gypsum, like alcohol a natural preservative; and by adding extra hops both in the recipe and after fermentation ("dry-hopping"). The malt used was not roasted black as for porter or stout, but left pale: hence, India pale ale.

Because hopvines twist clockwise to a length of twenty-five feet or more, they are grown on rows of poles strung with high wires and twine known as hopbines. The best growing districts, in England, Germany, Washington, and Oregon, produce distinctive types with names like East Kent Golding, Fuggles, Tetnang, Hallertauer, Cascade, and Willamette. English hops are dried in kilns shaped like beaked cones, called oasts. Much of the Cockney population used to evacuate London during late summer, as whole families took a working holiday picking hops in Kent. It was a laborer in the Northwest hopfields, inspired by the successful Campaign for Real Ale in the U.K, who launched the modern microbrewing movement in the U.S. Bert Grant worked to produce and improve *Humulus lupulus* for forty years, mostly for large brewers whose anonymous, watery products didn't deserve him. As a great authority on the subject, he knew he could make better beer than the gassy yellow norm. So in 1981 he launched Grant's Ales out of the old opera house in Yakima, Washington, the heart of hop (if not hip-hop) culture in the West. From this rivulet sprang a river of ales, all flowing into the ocean of microbrews we know today.

Though Bert is gone, the descendants of his well-hopped beers live on, and other brewmasters have taken his favorite herb to heights he never dreamed of. Beers with wonderful names like Hop Pocket Ale, Hop Ottin, and of course Hopalong all turn in formidable ratings as expressed in International Bitterness Units (IBUs). At the apogee is Stone's Ruination Ale, weighing in at 100 IBUs: "a liquid poem to the glory of the hop," reads its bottle. A Bud or a Coors, by contrast, would manage a mere 8 to 10 IBUs. I am lucky to have a couple of fine locals, the Fort George and Wet Dog, that field truly redoubtable IPAs; and a friend named Bob, holed way up a valley near here, who knows hops much better than I and uses them adroitly. My own personal favorite, however, for sublimity of both flavor and name, turned up on a late-night ramble to Lovejoy's, a highly hopped-up hippie alehouse at the unfashionable end of Sixth Street in Austin: Dennis Hopper Ale.

From all this hopgathering, my mind eventually returned to the question at hand: butterflies and hops. Indeed, several species employ hops happily as a caterpillar foodplant. The gray hairstreak (*Strymon melinus*), a pert mite of dove-gray below, graphite above, tomato-spotted on both sides, enjoys one of the broadest diets of any American butterfly, feeding on everything from cactus to cotton. Its larvae love hops so much that they sometimes become alleged pests in the hopfields. Then there is an eastern anglewing known as the comma *(Polygonia comma)*, which browses chiefly on elms and nettles. Like several close relatives, the comma noshes happily on *Humulus,* too—so much so that a lovely old name arose for it: the hop merchant. Finally, our common spring azures, the "sky-flakes" of Robert Frost's poem "Blue-Butterfly Day," have recently been split into a number of species with different ranges and food plants. One of these, *Celastrina humulus*, feeds exclusively on a native hop (*H. l. neomexicanus*) in Colorado gulches, and is known as the hops azure. In co-evolving with the bitter herb, perhaps the azures' larvae have incorporated its volatile oils as a self-defense against birds with a low tolerance for IBUs, just as monarchs use milkweeds' unpalatable white sap.

"Yes," I summed up, thirsty for a pint of the local IPA, "some butterflies are indeed fueled by hops." And, I might have added, so are most of the lepidopterists I know.

May/June 2003

Reality Check

When I was a boy, my family's escape routes from the suburbs to the mountains were any one of a dozen canyon roads into the Front Range west of Denver. Each one—Boulder Creek, Clear Creek, Coal Creek, Turkey Creek—had its own traits, which we came to know and watch for. Heading up Deer Creek, for example, leaving the weedy flats and cutting through the hogback into red hills of Gambel oaks, I always looked for the funny little sign that read: OLEO ACRES—ONE OF THE CHEAPER SPREADS. "Oleo" was short for oleomargarine, the recently concocted substitute for butter. The pun referred to early TV advertisements that maintained that "the cheaper spreads" would never match real butter's flavor or quality. But since we always used margarine, I didn't know any better. By the time I came to fully appreciate butter, cholesterol had come along, and it was too late to enjoy "the real thing" in good heart. (Though now we know that margarine, with its saturated fats, was much worse.)

So, what is real, anyway? A recent trip to Okinawa offered several opportunities to consider the question point-blank. From the first morning's gaze out my hotel window, I could see that Naha, the capital of the ancient Ryukyu Kingdom and of modem Okinawa, was of mostly recent construction. The Battle of Okinawa in 1944 had devastated the old city, leveling most of the wooden houses and coral-bedrock structures built over hundreds of years as if they'd grown out of the earth. Virtually all of the postwar buildings, strongly contrasting with the few surviving old houses, are made of cast concrete. Even the vast Shuri-jo Castle, a baroque confection of red-stained wood, coral limestone, and tile, has been rebuilt. The present skyline may or may not follow traditional architecture, but does this render it inauthentic? In his essay "Walking Downtown Naha," Gary Snyder described the newer buildings as being saved from the sameness one might expect in postwar concrete boxes by their varied shapes, colors, and lines; "they are jumbled, skewed, winding, tangential, and they rise and break with the rolling terrain," he writes—in other words, they are organic. Among the few vestiges of what came before, several stone pavements run steeply from the castle hill down into the warren of the old city. These paths cut a cross-section through uplifted coral cliffs whose eroded calcium carbonate is pocked with crannies, hollows, and cavelets stuffed with ferns and other limestone-loving plants. Concrete is another source of calcium carbonate, the soluble lime from which caves are carved by water and time. So it too becomes water-pocked, providing nooks for many of the same plants that colonize the coral walls. One particular plant, a vine called creeping fig (*Ficus pumila*), has tough, heart-shaped leaves that cling flat against the substrate and soften almost every hard surface in sight into a Cambodian or Mayan ruin. Along with the usual algae, lichens, and mosses seeking any place to grow, and the assiduous gardening by residents on every patch of exposed soil, these greens lend a living veneer to this city of cement. I came upon a tiny, ferny grotto behind our hotel, fashioned from both cut coral chunks and concrete, molded to receive and pool a little spring. Hunkering between the shiny hotel tower and a palm-and-cane-covered hill, it eased the passage from the manufactured to the growing, the human to the more-than, and sublimely demonstrated the relativity of "real nature."

It happens that Gary Snyder was also in Naha, where we were attending a conference of the Japanese branch of the Association for the Study of Literature and the Environment. In his keynote talk, he likened a working urban infrastructure to a state of ecological climax. Thinking of it that way, he said, gives a great opportunity for urban haiku. The urban and the rural, Snyder feels, are parts of the same phenomenal universe as the wild. Later, on a field trip to a historic site, I was scanning the off-coast rocks for birds. When Gary asked what I saw, I pointed out a gray heron perched on a mushroom-shaped remnant of reef. He glassed the peaceful scene. Then, grinning his impish poet's grin, he whipped off an instant haiku invoking the ancient castle site, the heron, and the tourist buses. The border between what's natural and what's not slipped another notch. For "natural," like "real," is often counterposed to "artificial," or "mechanical." Yet surely Snyder is right: nature includes the whole show, both oleo and butter, concrete and coral, birds and bushes and buses.

Reality Check

Still more dimensions to this question came out starkly during the conference. In my talk, I reiterated an idea I call the "extinction of experience," whereby the loss of common local species within people's easy reach leads to alienation, apathy, and further extinction—a particularly vicious cycle of disaffection and loss. A great naturalist and activist from Hokkaido, Mamoru Odajima, responded with a shocking story about children of his district who have long acquired and kept large stag beetles for pets. But as wild beetles have become scarce, they are more commonly bought than caught. Now, when the beetles die, the boys sometimes ask if they can get more batteries for their "toys." Well, why shouldn't they think the insects have simply broken down? After all, they didn't crawl through the brush to catch their beetles; and battery-driven puppies have been a big item here in recent years. Another speaker, Shoko Itoh, a scholar of American literature from Hiroshima, admitted that the great majority of children's free time is likely to be spent these days in what she called "the technosphere": on computer games, the Internet, and other electronically assisted activities. This being the case, she wondered, shouldn't we recognize the fact, and accord the technosphere its due place as experience worth having?

Maybe so. Virtual experience snaps the synapses just the same as the actual: *I feel, therefore I am.* And maybe that would be good enough, if it weren't for the fact that living systems desperately need saving, and few people care enough to conserve something they know only secondhand. But how much experience is enough? If you can't have wild nature, will captive do? Or how about the mere representation of the real? Our hotel in Naha posed a couple of piquant teasers along these lines. Inside the front door stood an endemic Okinawan shrub with shiny oblong leaves, hung with dozens of brilliant golden chrysalides the pupae of a big, black-and-white Asian relative of the monarch called the rice-paper butterfly *(Idea leuconoe)*. The hotel maintained a large butterfly cage where the larvae were reared on milkweed vines twined around trellises placed near the shrubs, so the wandering larvae would pupate on the ornamental bushes. These were then removed to the lobby as decorations, like lit-up Christmas trees. When the butterflies emerged a week or two later, they'd be released as part of an effort to re-establish the species in the city. Those who noticed the gilded pellets, as brilliant as polished drop earrings, were delighted and perhaps even inspired by them. In that same lobby, piped-in birdsong greeted me each morning when I came down to breakfast. This was pleasant. But would it move anyone, as the pallid thrushes and mauve violets behind the hotel moved me, to prowl farther along the wooded path?

Or how about the fish I saw at the end of my visit? While bidding goodbye to Gary, Korean poet Ko Un, and others in the Kansai airport in Osaka, I noticed an aquarium in the boarding area. Even captive fish can enliven an indoor wait, so I sidled up to the pretty fish and bubbles and ... they turned into an advertisement! The fish were virtual; the "aquarium," a cleverly designed video screen. Still they were riveting, until they vanished in favor of changing ads for Coke or Toyota, and the illusion was lost. The fish presumably were real, somewhere. But here and now, the fish, the bubbles, and the seaweed were merely excited pixels in an electronic tube. The phenomenon was real, all right, but the nature was nothing but bogus.

From the other side of the ocean, I can see that Okinawa broadened my notion of reality. But thanks to those airport fish, one of my deepest convictions remains intact: a bird in the bush is worth a thousand pictures.

July/August 2003

Guarding the Household

It was shortly before sunset on San Juan Island, the tail end of a bright May day in Puget Sound. Gulls plainted, semipalmated plovers piped, and a cool breeze came across the straits from the Olympics. Striding the tideline of Old Town Lagoon, I stepped from great white log to mussel-and-limpet-strewn cobble, from crispy green algae stretched like a drumhead to the coarse sand of the strand itself. The different textures clinked and crinkled underfoot until one step self-arrested in mid-air. There before me, right where my foot would have been, was a green-and-white butterfly roosting atop an orange fiddleneck flower. Scrunching down to its level, I gazed across a gray-green log to the pale blue tideflat, the deep blue bay, the black-green firs of Lopez Island, and again pale blue but the sky this time. All this, just backdrop to the orange-gold circlets of *Amsinkia,* its leaves the same yellow-green that marbled the linen-white wings of the very creature I had come here to find.

Until recently, this vision hadn't been seen since 1908. That's when the last specimen, of only seventeen known, was collected from the oak meadows of southern Vancouver Island—an ecosystem that had shrunk throughout the twentieth century as the city of Victoria grew. For nearly a century, this butterfly was considered extinct. Then, to the great surprise of Northwest naturalists, John Fleckenstein rediscovered the species some twenty miles and one country away, at the San Juan Island National Historic Park, while surveying grassland butterflies for the Washington Natural Heritage Program in 1998. At last it was given a name: *Euchloe ausonides insulanus,* the island marble. Viewing this one marble on its fiddleneck perch, secure from wind and tide among the driftwood, I thought about how life is a matter of place, time, and circumstance, all adding up to existence or erasure. By the grace of good luck, the butterfly before me represented a rare second chance at survival for a lifeform nearly lost.

To me, the word "homeland" conveys the sum of all such chances: the totality of habitat. So when I came in from the field that night and heard it taken in vain on the news, I realized what a travesty that holy term has once again become. In the department of instant clichés, there are few competitors of late for "homeland security"; and in the subsection of automatic oxymorons, there is no contest at all. The official rush to defend the country from threats external and internal, noble as the intent may be, has led to an almost complete absence of real vigilance, which must have to do with safeguarding the land where our home exists. "Homeland security" has become almost meaningless from sheer repetition in every conceivable, however inconceivably irrelevant, situation; and a cruel contradiction, denying by its own hoggish domination of the agenda any approach that might actually promote peaceful coexistence with other cultures and creatures.

Recently I picked up a newspaper full of dispatches from Iraq and the so-called Department of Homeland Security. Right beside them on the front page was an article headlined "Norton: No Additional Wilderness." The story explained that the Department of the Interior intends to cease review of its western land holdings for new wilderness protection, and to withdraw protected status from some three million acres in Utah. No isolated instance tucked away deep in the Great Basin, this action is typical of the many official attacks on our *actual homeland,* from proposing higher allowable levels of arsenic in drinking water to lowering power plant pollution standards, from increasing the cut of old-growth forests to declaring open season for oil and gas drilling on sensitive lands and waters. While Americans gird their collective loins with duct tape and seek enlightenment in yellow, orange, and red alerts, our leaders ignore the utterly essential prerequisite for any security that really counts in the long run: aggressive, passionate protection of the living and lovely landscape and waters upon which our very existence depends.

For what, really, is "homeland"? I like the description that appeared in the prospectus for a multimedia exhibit by that title at the Port Angeles Fine Arts Center: *As we move through the fog that clings to the slope, the homeland is always the beauty that's just underfoot.* Surely our search for a sense of deepest well-being must begin with the natural influences which Starbuck in *Moby Dick* invokes to "Stand by me, hold me, bind me." After all, a home is not a home without the land. I find myself again and again amazed by the seeming inability of governments to comprehend the absolute biological necessity of securing a healthy habitat before any other national or world priorities can make the least bit of sense.

Among all the news from the once-fertile crescent, how much have you heard about the Iraqi marshes? The reedbeds of the Tigris and Euphrates basins in southern Iraq, once larger than the Everglades, supported a remarkable culture of marsh dwellers living in uncommon harmony with the animals and plants of the waterways. Gavin Maxwell, in his 1957 book, *People of the Reeds,* described the marshes as "a wonderland," where "the colours had the brilliance and clarity of fine enamel" and "the reeds, golden as farmyard straw in the sunshine, towered out of water that was beetle-wing blue in the lee of the islands or ruffled where the wind found passage between them." Saddam's engineers drained most of the marshes in the early 1990s, depopulating them. But now comes a fervent effort, led by Iraqi expatriates, to rewater and repopulate at least part of the area. An expert commission has determined the goals of the campaign—named Eden Again—to be feasible.

Some places have not yet fallen from the grace with which they have been blessed by biology, history, and happenstance. Not really Edens, these countrysides—some wild, some suburban, some even urban—may or may not still function economically, but they are eminently productive of the diversity and wonder that lie at the intersection of nature and culture. Unlike the Iraqi marshes, they don't need heroic measures to be made whole again. They just need to be protected as they are, and to be managed to retain their rich mix of life and surprise. To save such treasures must be the purest of patriotic acts, just as the ultimate act of terrorism is surely the willful erasure of an evolved form or way of life.

How difficult this work can be when economic imperatives gang up against it, making conservation a task only Sisyphus could enjoy. But it is worth all the effort. For only in a world where all the bits still work—the watersheds, the pure air itself, the Swainson's thrushes that make of June evenings a madrigal of whistle, chitter, and tumbling trill; the trilliums in their soft, outrageous radiance in the green shade of cedars; the farms, forests, and homes, all in their proper places and proportions; the people, and all the other denizens of a world that needn't be lonely if only we acted as if it were really there—only in such a world may we feel safe. Surely the agency should be renamed as "Department of Habitat Security." For "Homeland" must include *all* the habitats, not just our own. And any state remotely worthy of the word "secure" will have to account not only for the safety of our buildings and their occupants, but for the safety of our nonhuman neighbors and their homes as well.

It is the island marble's good luck, and ours, that the place where it was rediscovered is already protected by the National Park Service. On my recent visit, I saw over one hundred individuals of this organism so recently thought to be extinct. Its future is not wholly secure, involving some tricky hostplant management challenges: the caterpillars feed primarily on adventitious alien mustards that park managers would like to eradicate. And of course, no species' existence is certain in the long run, while its extinction eventually is. But for now—for the time of our lives and for some period beyond—the island marble has a good chance, because of actions taken to safeguard and secure the very land where it makes its home. Will we be able to say the same for ourselves?

September/October 2003

A Tale of Two Turtles

That turtle didn't have a chance. Traffic was heavy on Maryland Route 16 not far east of the Chesapeake Bay Bridge: a long line of cars crowding behind me, a big truck bearing down from ahead. And there stood the turtle on stumpy legs, rearing its long neck over the centerline as if it were a finish tape. This was a big one—more than a foot long—and it was about to be smashed to bloody bits.

To me, the idea of a turtle with a broken shell has always been one of the more pathetic emanations of the chance universe, like snails crushed underfoot. Few sensations disturb me more than the soft, yielding crunch that means I've stepped on a snail. It's not the slime on my sole that bothers me. It's the sheer tragedy (albeit tiny) of a protected creature suddenly unhoused: still alive, but utterly vulnerable to beetle, thrush, and desiccating breeze—and what's worse, entirely beyond repair.

But broken turtles are even more pitiful, because they are long-lived vertebrates, presumably more capable of apprehending pain and distress. They can repair modest breaks (as can snails); but a badly shattered carapace means a slow, certain death. I think of the sad snapping turtle that John McPhee described in his great dining-on-roadkill essay, "Travels in Georgia." She'd been on her way to lay eggs when she was "run over like a manhole cover, probably with much the same sound." McPhee and his traveling companion, a canny herpetologist, had come across the tire-trodden snapper, "gravely wounded," but still alive and able to bite the hand that helped it. To put it out of its misery, they asked a passing lawman to shoot it with his service revolver. He did so, however ineptly. Then they respectfully cleaned, filleted, and consumed the unfortunate animal.

As a small child, I was passionate about two classes of objects: snails and suits of armor. Looking back, I wonder now whether the stresses of a sundering home—when such a state was rare—intensified my near-obsession with things whose purpose and metaphor was shelter. The fact that I dearly loved turtles too might reinforce such a theory; how often I wished for a shell into which I might snugly withdraw. A huge fossilized marine turtle in the Denver Museum of Natural History enthralled me more than the *Brontosaurus* towering above it, and I was crazy about pictures of Galapagos tortoises and loggerhead sea turtles in *Life* and the *Geographic*. The only chelonians I knew firsthand were the two-inch painted turtles we could buy at the dime store, then play with until they died from over-attention or neglect.

When I finally saw a live turtle in the wild, things did not go well. My brother Tom and I were visiting relatives in the Midwest with our grandfather in his big black Packard. Tom was thirteen, I was nine. An older cousin took us out walking on an Indiana woodland path. He put shotguns in our hands and said, "Shoot what moves." This is how cousins get shot. But this time what got shot was a turtle. A hand-sized painted turtle, much bigger than our ephemeral pets, surfaced from a pond near my feet. "Shoot, Bobby," my cousin commanded, "before it gets away!" I was transfixed. But I was a compliant child, and I pulled the trigger. Ever since, recalling that implacable blast, I have seen turtle all over the world. I bawled, and was disgraced.

Conservationists and animal lovers will always go 'round and 'round about the significance of individual lives versus populations. As one who works with insects, I am intimately familiar with these arguments and the conflicting feelings they arouse. The culling of individual insects very seldom has a significant impact on the population, and insect collecting is still essential for understanding and conserving diversity. Even so, when people spot me working with an insect net, they often ask in accusatory tones whether I catch and kill butterflies. Yet they don't give a second thought to using lawn chemicals, mosquito sprays, or bug zappers. Is a butterfly any more deserving of mercy than a mosquito? Or a moth? Or a mango? Where I live, most families hunt and eat elk and deer, but some of the best hunters I know couldn't harm a mouse. For my part, I readily admit to sentimental attachments that render me a distracted scientist at best. It's been many years since I could make a specimen of a red admirable butterfly or a mourning cloak.

So who's to say what we should do in a given circumstance of life or death? I remember several birding outings cut short by a dash to a bird rehab clinic with a disabled varied thrush or a sick scoter that died anyway. Yet, ruining a perfectly good insect net to catch an oiled murre off the Washington coast, rushing the mucky seabird to the spill-response field clinic, and following its progress to release, seemed the right—indeed the only—thing to do. I've watched waterfowl with their bills caught up in monofilament line, unable to assist them. And last week, on Cape May, New Jersey, an important staging ground for migrants about to cross Delaware Bay, I ogled a black-backed gull—biggest gull in the world—whose great right wing hung down. Alone on the beach, it stood uncomplaining (unlike many perfectly fit gulls), making little jerks with its wing, as if trying to sort things out; but it couldn't. My companion and host, a fine naturalist, explained that the local emphasis in this bird-famous place was toward conserving populations, not individuals—or on habitats, rather than rehab. Given available resources and needs, this made sense. Still, I took the look in that gull's eye with me when I left.

Roadkill, though, is a special category. Barry Lopez wrote movingly of honoring the deaths and bodies of highway victims in his essay "Apologia." I too have often been moved to place badger, flicker, owl, or raccoon on the soft grass verge rather than leaving them to be hammered into the asphalt; and last year, I dropped a broken-backed armadillo into a moonlit Arkansas creek, reckoning drowning to be the quicker death. I've also tried to respect, if not redeem, the dreadful toll of our easy mobility by taking the dead—an otter, a heron, several hawks, and many smaller creepers and fliers—to zoology museums. When yearling deer die on the highway above our house, we wheelbarrow them down to the field below, open their sides for the ravens, and watch the succession of scavengers return them to the flow. And earlier this spring, Thea and I paid tribute to a California quail cock's untimely demise against our bumper with offerings of fresh morels and wild asparagus foraged that same day, the whole prepared and sanctified with a decent merlot.

Ah, Death, what is the deal with thy sting? We each take plant and animal life, whether intentionally or not, in order to live. We each have our biases, and we all indulge them. You could say that "kill and let live" is the human way. For that turtle on the highway, pondering such questions was a luxury it couldn't afford.

I braked the rental car in the middle of the road, stabbed on the safety blinkers, threw open the door, leapt out, and put my hands up both ways, like some cartoon traffic cop without a whistle. This is how good Samaritans die. But everyone stopped, so I ran toward the turtle. "Please, Pan," I mouthed, "don't let it be a snapper!" Pan smiled; it was a great big red-belly. As soon as it felt my footfalls, the turtle pulled its legs and head deep between its carapace and plastron. I reached down and plucked it up from the road, grasping both sides of its heavy shell. No one even honked.

November/December 2003

The Chemistry Between Us

As a boy, I subscribed to a monthly mailing called "THINGS of Science." Each time the blue cardboard box landed in our mailbox, I'd eagerly turn back the copper clips and lift the top to see what was inside. I hoped it would contain something to do with plants, bugs, or shells, rather than some boring experiment in chemistry or physics. Likewise, I neglected our home chemistry set in favor of chasing butterflies. Inanimate stuff just didn't ring my bells with the same sweet timbre as the living. So even though my high school biology class with the football coach, Mr. Buchkowski, was no great shakes, chemistry came as a definite step down. Take the didactic drone of Mr. Blubaugh, he of the obvious nickname; add my narcoleptic afternoons amid the acrid smells from something nasty on the Bunsen burner; and you had a red-faced chemistry teacher in full roar. "Wake up, Pyle!"

A couple of years later, as a would-be zoology major at the University of Washington, I was again imprisoned in chemistry class when I could have been out birdwatching on the campus marsh. I often forsook the sour reek of Bagley Hall for the arboretum, and the result almost scotched my college career. I consoled myself for the rotten grades by concluding that chemistry is balderdash. At least I'd learned my birds.

In the soundtrack of my childhood, "Better Living Through Chemistry" was a frequent mantra. After all, we were the leisured and lucky recipients of countless postwar chemical boons: nylon, Scotch tape, and Melmac, to name just a few. No one, it seemed, doubted the rosy future both promised by and chock full of lots of lovely chemicals. Until Rachel Carson. When I read *Silent Spring,* I learned for the first time how very ironic the slogan— "Better Living Through Chemistry"—really was. The lawn where I'd passed countless childhood hours was free of weedkillers, except for my brother and me, digging dandelions for a nickel a peach-can-full. No insect spray either; the grass hopped with tawny-edged skipper butterflies as well as kids. But Rachel told another story, one I would come to learn firsthand,

91

as Denver lawns became skipper-free biocide sponges under the influence of ChemLawn, Monsanto, and Ortho: places where turning kids out to play should be considered a form of child abuse.

But chemicals themselves aren't responsible for what we do with them. They are, in fact, the basis of all life, and everything else as well. By dodging my chemistry classes, I'd undercut my education. A little late, I realized that natural history is much more fascinating when one knows something of the underlying chemical processes. Learning how insects take in plant compounds to render themselves distasteful to birds helped me to see this. Plants and insects coevolving to the advantage of each include the age-old dance of white butterflies and mustard glucosinolates; the partnership between cinnabar moths and the pyrrolizadine alkaloids borne by their tansy ragwort hosts; and the well-known relationship between monarch butterflies and milkweeds with their cardiac glycocides. Experiments proving that birds learn to avoid unpalatable monarchs gave rise to the whole field of chemical ecology, eternally and unforgettably symbolized by Lincoln Brower's iconic photograph of the barfing bluejay in the February 1969 *Scientific American*. Chemistry, I realized at last, could be fun.

Then I discovered that the substance of the land itself provides another natural link—chemistry made manifest in plants and butterflies that occur only on certain substrates. Rare orchids and blues that may be found only on chalk grasslands in England, for example, or cobra lilies and skippers restricted to serpentine outcrops in northern California, and others that have learned to thrive on toxic nickel deposits in Washington's Wenatchee Mountains. This is indeed a world made equally of *chemo* and *bios*. The fact is, as the Greek philosophers well knew, mineral and organism are merely ends of a common chemical continuum.

A more recent chemistry lesson struck closer to home. In August, my wife, Thea, abruptly learned that she had ovarian cancer. Since then, she's been receiving a compound of platinum in a vein near her heart every three weeks, because platinum was once accidentally observed to prevent the growth of cancer cells. Coincidentally, the day before her first infusion, I too received the same element: two crowns and an intervening bridge, contrived of porcelain, platinum, and gold. The platinum, it turns out, has thermal qualities that allow the porcelain to adhere to the gold without cracking. In a strange way, this all hearkens back to Mr. Blubaugh. I recall the big periodic table of the elements that filled much of the front wall above the blackboard, and how I'd try to stay awake by conjuring on the properties of the various elements, which we had to memorize. I remember being surprised to learn that there was a metal more precious than gold. Elvis's records all went gold; but by the time the Beatles swept the charts, the very best-selling singles were going platinum. Still later, platinum credit cards superceded gold cards as the most prestigious plastic. I was nicely disabused of any notion that this actually meant anything when my own Visa card perfunctorily mutated from ordinary to platinum without passing gold. But now, platinum has taken on a whole new meaning for me—coming into our lives (and beings) in much more direct and visceral ways.

Soon after our dual lessons in the properties of platinum, I heard on the radio that it is now twice as valuable as gold, nearing $800 per ounce. Platinum is mined chiefly in South Africa and Russia, and used mostly for jewelry and in catalytic converters for cars. The story didn't mention oncological uses, but thousands of women owe their lengthened lives at least in part to Carboplatin. Thea also receives taxol, another potent anticancer drug, discovered in yew trees. Many wild yews were cut for their bark before chemists learned to synthesize taxol from yew needles; and the winning and refining of platinum is not much gentler on the land than the notoriously toxic methods of mining for gold. We know that Thea's salvation exacts costs beyond our medical bills. So it is when we dabble in the great chemistry set of life.

I recall that Rachel Carson died of cancer, even as she fought to alert us to the fatal dangers of chemical pollution. Now she looks over our shoulders and rolls in her grave as we go on living in a world so thoroughly invaded by synthetic chemical compounds that the toxic burden in our bodies is enormous and the physiological effects profound. Almost all of us have detectable Teflon in our tissues. Breast milk commonly contains DDT, PCBs, and mercury. Organophosphates from insecticides lurk in our livers and infiltrate our fat. Dioxins, the villains in Agent Orange, lace our bodies and waterways. Glyphosate, the active ingredient in the most common weedkillers, has been implicated in lymphomas and various reproductive ills. Cancer swarms flare in agricultural areas, industrial zones, and in the shadows of nuclear test and waste-disposal sites. Remember how rare and unspoken cancer used to be? Now, every family seems to be affected by one cancer or another, and words like "biopsy," "chemo," and "radiation" are as common as aspirin. Breast cancer alone can be called nothing less than a plague. Conscionable chemists predicted long ago that the baby boomers would be the generation to reap the carcinogenic legacy of the Century of Progress. And so it is.

So is chemistry balderdash, or the source of better living? Now, with my love's life in the balance, I know it is both. Just as monarch butterflies have evolved the ability to sequester cardenolides and turn them to their own advantage, we internalize expensive potions like platinum—in our blood, in our mouths, in our cars—and adapt their qualities to our own ends; those of us, that is, with the platinum in our wallets to pay for it. Our Faustian tradeoff seems to be to live in a world rendered so hostile by our chemical excesses that we can continue to live only by extracting, refining, testing, and ingesting still more chemicals.

Mr. Blubaugh, wherever you are, you can relax now. I'm finally paying attention.

January/February 2004

NOTE: This was written in 2004. As of 2012, after many ups and downs, Thea is thriving. Now in her eighth cycle of chemotherapy, she has become allergic to Carboplatin and resistant to other chemicals, but newer taxol-based potions are proving very helpful. We both harbor deep thanks for the loving care and concern sent by so many readers since this column originally appeared.

Taking Their Names in Vain

"You're so sluggish!" Thus spake my lively mate one recent morning, when my late-night reading and early a.m. torpidity conspired to bring about yet another tardy departure for the city. Of course she was right, and her frustration was righteous. Nor was her metaphor inappropriate. Like our damp land's wondrous banana slugs, I function on average at a more deliberate pace than her pet rabbit, though I (like the slugs) am actually capable of a speedier slither now and then. On the whole I was not offended by the comparison.

People have always described one another's traits in terms borrowed from the bestiary: sly as a fox, quiet as a mouse, blind as a bat, slow as a turtle (is that slower or faster than a slug, I wonder?). Meddlesome or raucous or clever as a jay or a crow or a magpie. We've got hawks and doves, WASPS and crabs, shrews and moles. Animal similes that anthropomorphize animals and zoomorphize people have been used and used and will be used again to portray our own species' qualities. Never mind whether the attributions fit. True enough, bees and beavers are pretty busy, and bugs in rugs are probably snug. But are eagles actually brave? Owls truly wise? Geese really stupid?

When such name calling is meant as an insult, it may carry an ironic compliment. Calling your uncle a turkey, you have in mind a butterball of an inbred, domestic version gobbling inanely as it bumbles about the barnyard brainlessly, and the likeness might not be inapt. But when I saw a sleek, bronze-backed, Crayola-tailed wild turkey pacing cannily beside the Blue Ridge last fall, the insult lost its punch. As all hunters of wild turkeys know, there is nothing bumbling or dumb about their quarry. I think too of the three-toed sloth I once watched in all its slo-mo elegance, negotiating the canopy of a Costa Rican rainforest. Here was an organism so exquisitely suited to its domain that it scarcely needed to move, supporting an endemic flora and fauna in its dense, lichened, gray-green coat. Yet to most, the word "sloth" evokes nothing more than one lazy louse.

Our animal pejoratives begin gently enough, such as catty and bird-brained, but then deteriorate through dirty rat and hairy ape and on to louse, skunk, hog, and horse's ass, peaking out at just plain "animal." Before long we're in the territory of genuine slurs—women as bitches, chauvinist men and police as pigs, anyone in the wrong tribe as running dogs—never mind that we consider all three categories honorable when describing Lassie, Porky, or Old Yeller. Slander is bad wherever it occurs, but to co-opt animal names in its service leaves me doubly uncomfortable; not because of the nasty animal attributes such names are meant to imply about the people thus anointed, but because of the nasty human traits with which they tar the innocent creatures concerned. By describing insufferable hangers-on as leeches, ciphers and sycophants as toadies, craven cowards as chickens, and liars as snakes in the grass, it is the animals we really slur. Through no fault of their own, critters end up in the kind of company we'd prefer to avoid.

When the first U.S.-Iraq war was underway, I noted that a well-known John Bircher's billboard along 1-5 concluded its slogan of the week by labeling Saddam Hussein "a human worm." The worst aspersion that farmer could think to cast was the name of Charles Darwin's darling: "It may be doubted," Darwin wrote, "whether there are many other animals which have played so important a part in the history of the world." During the second invasion, an interviewee on the National Public Radio program "Fresh Air" referred to Saddam as "a lying snake," and once caught, he became "a rat in a hole." But this sort of infamy by association cuts both ways. As momentum built for George W.'s Iraqi adventure, a San Francisco columnist described the administration as "vile power-mad slugs and lizards ...snakes and pit vipers." The writer went on to admonish his readers, "The world does not consist of simpleminded and reductive good/evil polarities, but, rather, is a living organism, interconnected and breathing and dying and renewing in constant flux." But what manner of Gaia is this, where all sorts of creeping, crawling creatures are fit only to characterize our dictators and demagogues?

When the worst thing you can think to say about someone—who may in fact be quite vile—is that he or she is an animal of some kind (never mind whatever mix of precisely evolved, adaptive, miraculous traits that creature in fact possesses), then biophobia is in full bloom. And biophobia, however unthinking, has its consequences. This sort of negative taxonomy sets us up to do some real damage, for as soon as we derogate life forms by tying them to our own worst side, it becomes easy to objectify and even obliterate the subject of the insult. An Internet commentator recently referred to George W. Bush as "the Smirking Chimp." How does such a put-down affect our attitude and responsibility toward actual chimpanzees suffering from habitat loss and the bush-meat trade in Africa?

I am always baffled by folks who claim to love the Lord and all His works, and then go on to damn any endangered species they see as goring their own ox. Yet surely the knowing annihilation of any product of the Creation—by any creation myth, Darwinian, Judeo-Christian, Algonquian, Aboriginal—is deeply, shamefully sinful. Could it be that by dragging animals down to our own level, we can justify treating them just as badly as we treat one another, and even worse? After all, we place people in jails, and slavery has an ancient pedigree, so should it surprise us when we confine elephants and leopards in cages, turn asses and camels into beasts of burden, and cattle and sheep into vehicles of meat? Medical and nuclear military "experiments" on humans and all manner of torture have much in common with vivisection and other laboratory abuse of rats, rabbits, dogs, and primates. And why, having eliminated numberless tribes and nations of indigenous people outright, should we scruple to prevent the extinction of entire runs of salmon or herds of caribou? Who is going to mourn the loss of endangered suckers and bullheads in the Colorado River, when we use those words for people we find beneath contempt? The fact is, when we zoomorphize people, we make it all the easier to behave inhumanely toward the objects of comparison.

Zora Neal Hurston once explained the African-American penchant for Uncle Remus-type animal stories by saying that "we throw the cloak of our shortcomings over the monkey." She had that right. And there is nothing new about this practice—you'll find it in the Bible, you'll spot instances in Shakespeare, and you'll certainly see it throughout both Eastern and Western fables, not to mention on billboards and talk shows and all through the common parlance in the market, mall, and chat room. Oh, yeah, we'll always be scared as a rabbit, hungry as a hyena, horny as a horn-toad. But the next time it occurs to you to put people down royally by calling them creatures, consider this: how would the snake feel about it? How about the weasel?

I can appreciate these creatures' plight. After all, I know what it's like to have one's name taken in vain, having grown up with my particular surname. I'm just glad it isn't plural, and that my grandfather had the good sense to change the "i" to a "y." Still, when conversation comes 'round to piles of things, I'd like to say fine, stick with your piles of books, peaches, leaves, and compost, all good company; but when it comes to scat, can't we please say "heaps" or "mounds"? You'd know how I feel if your name was a collective noun. By comparison, "sluggish" is a treat. Only I wonder: do the slugs appreciate the likeness?

March/April 2004

Tit for Tat

Amtrak Cascades train #510, northbound for a holiday getaway, Seattle to Vancouver, B.C. Morning alpenglow pink on the entire eastern rampart of the Olympics. Puget Sound plashing trackside, and just yards from the freshly washed windows, harlequin ducks. How beautiful the drakes, all rust and gunbarrel blue, clean striped and sickled with black and white. How utterly indifferent they were to our wowed attention. Even the rushing locomotive, similarly striped in cream, brown, and teal green, caused them no distraction from their winter morning's forage. We looked at the harlequins, but they didn't look back.

The following day, we rounded Beaver Lake in wild-in-the-city Stanley Park. A wet snow was falling, and wildlife thronged the path—spotted towhees, varied thrushes, and both the black and the grizzled forms of the introduced eastern gray squirrels. Most of the creatures, well conditioned to begging by walkers free with foodstuffs, were importuning us. Then a pack of little birds appeared in the salmonberry beside the path. I first thought bushtits, but their buzzy deeee-dee-dee and the habitat, more coniferous than deciduous, gave them away as chestnut-backed chickadees, the most colorful of North American titmice. Their mobile mantles of fluff glowed the same rich russet that graced the harlequins. But unlike the ducks, these birds were anything but aloof; they clearly sought our attentions. We had nothing for them, and anyway, feeding wild animals is seldom a favor in the long run. Helping snowbirds and stay-behinds through a killing cold snap is one thing, but helping to habituate trailside dependence is another.

Nonetheless, I saw the opportunity that presented itself, removed my glove, and raised my left hand. In a feather's flutter, a C-B C-dee flapped to my finger and remained for a second or two; then another did the same. At one point three of the tits flirted with three different fingers at once. I knew it wasn't fair. I was exploiting their hopes of a free lunch to get a cheap thrill. But I did it shamelessly, for some time, and was deeply charmed. I think I will always be able to see those black onyx eyelets in the ebony cap, to feel the prickly tingle of the teensy talons on my fingertips. Pure magic.

What do you suppose the chickadees got out of this encounter? Sheer frustration at unmet gratification of a learned behavior, I'll bet; nothing more. And yet, we all want more than that from our natural encounters. People seek reciprocity wherever they can get it, and many places they cannot. I have noticed that many ardent nature lovers take it as an article of faith that their beneficence bounces back: sort of the pantheist's version of "Jesus Loves Me." Then when they have a bad day with mosquitoes, they feel spurned. When we hug a tree, does the tree hug back? Or is our love of nature unrequited? We all wish for two-way traffic on the highway of life, but does it really work that way? Bison biologist Dale Lott considers this desire to be "part of our romantic illusion about other animals," and he doubts that they "reciprocate our tender regard or much of anything else." But does it matter, as long as we imagine that nature cares about us?

Bennett Cerf, the great humorist and founder of Random House, liked to tell a hoary joke about a wizened old bachelor who decided he should have some company at home, so he attended an auction for a parrot. The bidding started low, but each time he bid, someone raised him until the price grew much higher than he'd ever intended to pay. When he got home, he commanded the parrot to speak. Silence. Again he called for some company; again, silence. Finally, the man erupted: "Blast! I paid a thousand dollars for a parrot who can't talk?"

Can't talk? squawked the parrot perfectly. *Who do you think bid you up to a thousand bucks?* Of course, the question of whether a parrot has anything to say or is merely parroting its trainer is an old one, and in modem times the latter interpretation has prevailed. But lately, some ethologists have wondered whether the answer is quite that simple. Few people expect colloquy with a goldfish, but we all believe beyond a doubt that we experience emotional exchange with our cats and dogs. And when it comes to chimps, the work of Jane Goodall, Roger and Deborah Fouts, and others makes it quite clear that a genuine exchange of thought and feeling takes place. And why should it be otherwise, between ourselves and our closest living relatives?

As in most other so-called dualities, a continuum connects these two states: sapient response did not evolve whole cloth with *Homo sapiens*. So while we may be justified in supposing that most insect behaviors toward us are impersonal, red admirables picking out the same person to alight on over and over and monarchs making decisions during migration really make you wonder. And no one reads Gavin Maxwell's *Ring of Bright Water* and imagines that the otters he keeps (or that keep him) do not engage in lively, thoughtful interchange with the author. It's a long way from a Skinner box to a meaningful relationship; but at what point does conditioned response become affection, curiosity, or regard, apart from any reward?

Even if we imagine ourselves going hand-in-paw with the animals, we would be wrong to expect all such encounters to be pleasant ones. Perhaps we should be careful what we wish for. After all, many of the face-to-faces between humans and large animals, throughout much of our history and prehistory, ended with someone eating someone else's face. Hunting cultures often praise, beseech, and honor their quarry; but when humans become the hunted, reciprocity takes on a whole different tang. David Quammen's tour de force, *Monster of God: The Man-Eating Predator in the Jungles of History and the Mind,* tells in visceral detail just how far this particular sort of one-on-one has gone. I wonder if anyone has ever submitted happily to such a fate, as hunters bid deer and salmon do for them? Now, that would be unconditional love.

I've experienced just one scrape with actual predation, but it was definitely enough to make me reconsider the desirability of give and take with all that creeps. Canoeing in the bayous of the Trinity River near Houston, I was peacefully paddling bow in the lead canoe, bewitched by the dwarfish forms of the bald cypress knees and the Gothic arches of their buttresses.

And then ... Wham! I was a foot out of the water, leaning at an angle that felt like an imminent roll, screaming an epithet that furnished Our Saviour with a whole new middle name. There was the eye of the most massive alligator I have ever seen—no more than eighteen inches from my face. Happily, my sternsman was expert at keeping a canoe upright, and he maneuvered with his paddle to do so as we slammed back down and the 'gator slithered slowly out from under us in the mud. Then it burst into deeper water and we watched it power off upstream, the full twelve feet of it visible until it dove. Had we capsized, I would have been dumped right onto the giant herp's back. Its response, if not to twist and bite, would have been at least a mighty tail-thwack. With a nervous laugh and an expletive of his own, my swamp-rat host pronounced that I was indeed a lucky man.

On the whole, I think I'll stick with chickadees. And recalling their impatience with my extended but barren digits, I am driven to ask: Is this a fair way to regard the world and its creatures, to ask what's in it for us? None of us wishes to be out on this great big limb alone, and sometimes a magic moment actually allows us to imagine otherwise. But it might be that the best way to love nature is without expectations. If we can accept that the world doesn't actually give a fig about us, and love it just the same, then maybe we're getting somewhere worth being.

May/June 2004

Small Mercies

Between the Space Needle and Puget Sound, among tree ferns and bromeliads, we watched scads of neotropical longwings, massive blue morphos and banana-sucking owls, little scarlet-and-black swallowtails, aqua-neon preponas, and key-lime-and-opal malachites. My wife, Thea, was about to have surgery, and these were the images she wanted to carry with her into the fog of anesthesia.

The butterfly house at the Pacific Science Center is a typical butterfly conservatory, where tropical conditions prevail under glass as in a Victorian hothouse, allowing one to walk among exotic species during the northern winter. Immensely popular, butterfly houses have sprung up in many cities. Children on school break thronged this one. All visitors had been cautioned on the way in not to pick up the butterflies, and especially not to step on any that may have alighted on the paths. But one little lad who hadn't quite gotten the message spotted a zebra longwing at his feet, and drew up his sneaker-clad foot to stomp it good. Happily, his mom stopped him just in time, and gently informed him that you don't tromp on the butterflies in a butterfly house.

Where did that child get the impulse to crush? There is always plenty of entomophobia going around, but most kids have a native fascination with bugs. To just up and smush it like that, you'd think he must have had a pretty direct example. Some parents off every interloping insect, setting a destructive pattern, though obviously not that watchful mom. Maybe the misguided tyke had an evil babysitter with a particular vendetta against insects?

Later I picked up a *Seattle Times*. When I'm in the city I try to see it, if only to follow the weird doings of that subtle, rasty cat named Bucky in the comic strip "Get Fuzzy." To get to Bucky, you must pass the top-billing space on the funnies page, where resides that other cantankerous feline of venerable standing in the comics community ... and there was yet another instance of Garfield gleefully smashing a spider with a rolled-up newspaper. The spider says, "You can swat me, but there will be another spider to take my place." "Very well then," replies the cat. SMACK! "I'll renew my newspaper subscription." As if that wasn't enough, the next day's installment showed a spider hanging back, saying, "I'm not coming any closer." SMACK! "My latest invention," smirks Garfield. "Magazine on a stick."

Butterflies and spiders are scarcely interchangeable. But a longwing is leggy, and with its wings folded, it might well elicit an aggressive response of the sort usually reserved for unwelcome insects and spiders. I wonder whether that child, or any other readers, may have been inspired to commit random acts of violence toward inoffensive creatures by the misarachnistic behavior of Jim Davis's famous cartoon kitty. After all, Garfield gets around; from Garfield boxer shorts to suction-cup Garfields in car windows, this puss commands a tremendous sphere of influence.

Which takes me back to a recent autumn, traversing the Midwest with a group of writers. Outside Muncie, Indiana, we were cordially received at Jim Davis's farm and compound. The cartoonist was not actually present, but his staff showed us various progressive, conservation-minded features of the place: a state-of-the-art biosewage treatment system, where wastewater was processed by wetland plants to remove impurities and bacteria and return clean water and valuable nutrients to the ecosystem; and a prairie restoration program, remodeling farmed-out weedy fields into a more natural state.

The autumn fields bore rich raiment of purple asters and yellow goldenrod, nectaring thirteen species of butterflies—a lot for so late in the year. No morphos or malachites, but meadow fritillaries, orange sulphurs, and eastern tailed blues flashed over the meadow. No wonder we left the region regarding Garfield's dad as a sensitive environmental innovator. Hence my confusion whenever the spidey-smashing panels appear—which is at least as often as the snake-smashing sequences in "B.C." (Johnny Hart, this goes for you, too!)

Now some may say that this is just cartoon violence, like Wile E. Coyote getting bonked, blown up, or flattened. Maybe; and maybe kids and parents are smart enough to realize this, and to recognize irony. But there is no irony in Garfield's repetitive acts of carnage—spider appears, cat whaps spider, end of story. Lesson: When anything creeps or crawls into your space, smash the hell out of it. Between Raid and general scapegoatery, the so-called "lower" forms of life have enough going against them that they really don't need this too. It's not a matter of pest control or self-defense. Very few spiders pose any threat whatever, most putative "spider bites" are nothing of the kind, and not even black widows or brown recluses leap out of nowhere to eat our faces or bite our butts. Few spiders become economic pests by competing with us for food or fiber. Bopping arachnids at random doesn't do anyone any good, and may actually increase pestiferous insects by knocking off some of their most effective predators.

A recent world conference at the American Museum of Natural History, called Expanding the Ark, confirmed that small-scale lifeforms comprise the foundations of life's pyramid. When they fail, such as pollinators on land or krill in the ocean, the rest come tumbling down. Arthropods account for most of life on Earth, and most of them are insects and spiders. Most species are benign or beneficial to us, and essential in food webs. That genuine, functioning prairie that Mr. Davis seeks to restore would support millions of spiders per acre, belonging to hundreds of species. And watched closely, they would reveal intimate beauties, exquisite form, elegant function, and astonishing stories of adaptation. In case any Garfield readers stand in need of remediation on these points, I highly recommend membership in the Xerces Society for invertebrate conservation, and a subscription to their beautiful magazine, *Wings*. Be sure to request the glorious recent issue on spiders. Or if it is merely a matter of arachnophobia on your part, let me reassure you that I was the worst of arachnophobes at a young age—I couldn't even approach a field with a garden spider in residence without screaming. This was a serious handicap for a young lepidopterist, and eventually I overcame it completely. We are all educable.

I hate to be a scold, when I otherwise enjoy Mr. Davis's strip. But when I think about the message he may be sending to would-be arachnophobes, I am saddened. Will generations of young spider smashers endanger any species? No. But gratuitous acts of killing innocent creatures are simply not a good thing to model, nor are they funny. And it cuts both ways: As entomologist/writer Eric Eaton puts it, "If more people felt honored to host a diversity of living things in their own homes, sheds, yards, gardens, and garages, then there would be a collectively different attitude toward biodiversity in general, one that embraced all manner of creatures." But can we hope in good conscience for a world of respect for living things when one of our premier cultural icons smacks defenseless spiders on sight, over and over? Not that I'm putting Garfield up as a role model for the young, his other main character traits being gluttony and sloth ... well, maybe he's been more of a role model for Americans than I thought!

Mr. Davis, if you are reading this, listen up: You communicate with more people on a regular basis than any politician or social leader, and you do set an example. I invite you to honor your principles and stature as a conservationist, at the cost of rendering your feline alter ego slightly less irascible. Please use your power to strike a blow against unnecessary cruelty and biophobia. The next time your portly pussycat meets a spider, have him show it the door, with a smile instead of a swat. After all, even rasty cats have karma.

July/August 2004

NOTE: *All manner of erroneous spider assumptions are dispelled in an authoritative and entertaining manner at the Web site of my arachnologist friend, Rod Crawford: http://www.burkemuseum.org/spidermyth/*

The Long Haul

In the dim deepwood of massive and moss-bound trees, the three tenors of the Northwest forest give voice: varied thrush's raspy note, like whistling through spit; golden-crowned kinglet's high tinkle, the sound older ears lose first; and winter wren, puck with a pennywhistle on an endless tape loop. A fourth, pileated woodpecker, is silent for now, having already totemed all the big old snags.

I've arrived at a place known as the Log Decomposition Plot. The mossy turnoff is paved in evergreen violets, then comes a trench and berm to keep vehicles out, but the bulldozed tank-trap has grown to resemble a native outcrop, covered in sword fern, salal, and moss. Fresh wind-throw renders the trail almost impassable at times: a suitable gateway to a place where, when a tree falls in the forest, a lot of people hear it—and then take a close look at what happens next.

When I get to the laid-out logs and the sawed-off tree-rounds that fallers call cookies, I know I've arrived at the place where druids of forest research make offerings to Rot. This is the H. J. Andrews Experimental Forest, 16,000 acres situated deep in the Oregon Cascades and managed by Oregon State University and the U.S. Forest Service. The Andrews, dedicated to forest research since 1948, became a charter member of the National Science Foundation's Long-Term Ecological Research Program in 1980, one of twenty two sites in the United States and two in Antarctica. The fundamental study of the northern spotted owl took place here, along with much basic research on forest function. Recently, recognizing that science is not the only tool for probing what forests mean, the Forest Service and the Spring Creek Project of OSU's Philosophy Department initiated a program called Long-Term Ecological Reflection. This inspired whim is the source of my good luck in spending a week here, reflecting and writing.

Whole watersheds of old-growth western hemlocks and Douglas-firs that grace the Andrews are simply shocking compared to the second-

and third-growth evergreens of my home hills. The Decomposition Plot, devoted to studies of nutrient cycling and forest refreshment, lies in one such ancient stand. It's easy to tell when I'm inside the research zone by the yellow, red, and blue tags on wire stems sprouting from the moss. One pink cluster pokes like old trilliums from a mossy mound that once was a tree. A red bunch limns the ground where a one-time log has finally given up the ghost. Metal tags label the cut butt-ends of many logs that lie about higgledy piggledy, as gravity and the wind might have arranged them had researchers not dropped them first. Bright flags beribbon trees, shrubs, small boles, and limbs, and duct tape shores up the ends of some logs: is someone investigating the degradation rate of duct tape as well as wood fiber? White plastic pipes, buckets, jugs, and other bits lie here and there, each significant to some experiment or other. In early spring, no one is here for me to ask.

Some would see all these artifacts as litter, marring their wilderness experience. You can also see them as inflorescences, like that mysterious white plastic funnel sprouting next to a nodding trillium. Take away the pink ribbon around that hemlock over there, pick up all the aluminum and plastic, and this old-growth forest would still work just like any other. Researchers cut fresh cookies for a starting point, then measure their decay forever after—or as long as they can. But let all the straight cuts rot away and you've got an untidy place going about the important business of trading in the old for the new, an ecosystem definitely in it for the long haul.

For the most part, most of us take the short-term view, most of the time. What gratifies right now, or soon at the latest, is always more compelling than what might satisfy years from now, let alone nourish the generations. When business opts for short-term profits instead of long-term husbandry, both forest and human communities suffer. The short view is what turned most of the Northwest's giant forests into doghair conifer plantations cut on short rotation for pulp. To peer much further down the line requires not only empathy for those who follow, but also faith in the future—even if you won't be there to see it for yourself. Such an ethic underlies all of the long-term studies here on the Andrews, whether concerned with old-growth ecology, hydrology, riparian restoration, forest development and mortality, carbon dynamics, invertebrate diversity, or climate change and its effects.

Meanwhile, here in the Decomp Plot, nuthatches toot in monolithic columns of Douglas-fir; a robin chitters in a clearing. Dappled light falls on forests of the moss called *Hylocomium splendens*, hammocks of shiny twinflower leaves, and fleshy *Lobaria* lichens lying about like tossed-up ocean foam. The path is a maze of Irish byways for voles. Douglas squirrels leave their middens of Douglas-fir cone bracts all about like a prodigal's spent treasures, and round leaves of evergreen violets and wild ginger spatter the path like green coins. If they were gold, I doubt they'd distract the unseen leprechauns who come here to gather the data of decline. Gold doesn't decompose, and this place is all about the documentation of rot. It goes on all around me: something fairly large just fell from a nearby old-growth giant.

Maybe that's the problem with the long view: it speaks of our own inevitable demise. We're not much into self-recycling. Even in death, we take heroic steps to forestall rot by boxing our leavings in expensive, hermetic containers. After all, to anticipate the future—a future without us—is asking quite a lot. But life and regeneration are the name of the game on this mortal plane, every bit as much as corruption. The winter wren's song, after all, is no morbid message. Old vine maples hoop and droop under their epiphytic shawls, but the unfurling leaves of the young ones are the brightest items in the forest (even brighter than the red plastic tags). Every downed and decaying cylinder of cellulose makes yards of nitrogen-rich surface area for hopeful baby hemlocks, lichens, liverworts, and entire empires of moss to take hold on and begin making forest anew.

If we care about what's to come, it makes sense to send delegates to the forests of the present to find out how things truly are, report back, and check in again year after year. The conundrum of the diminishing baseline says that if we have no clear idea of what went before, we are more likely to accept things as we find them, no matter how degraded they may be. Memory is short, the collective memory even shorter. But with baseline in hand, we can appreciate change for what it is. Recognizing loss, we may even act to prevent future loss.

Just as the scientists gather data, any open-eyed observer could go on documenting details without end in such a place: the declination of that row of saplings bent over one deadfall by another; the way that one sword fern catches the sun to suggest a helmet; how the polypore conks launch out from cut ends as soon as they can after their vertical hosts go horizontal, their mycelia reorienting ninety degrees to the zenith. There is no end to particulars as long as the forest goes on and there is someone to record them. The moss grows, the raven barks, the trees go to soil—first hemlocks, then firs, finally cedar. All the while, the decomp team is there, watching how the cookies crumble. Maybe looking to the future is a way of hoping there will still be something to see when we get there. Maybe it's the only way to make sure of it.

September/October 2004

NOTE: *Long-Term Ecological Reflections is a collaboration between the Spring Creek Project for Ideas, Nature and the Written Word at Oregon State University, the Andrews Forest Long-Term Ecological Research Program, and the Pacific Northwest Research Station, with funding from the US Forest Service. It was the brain-child of Kathleen Dean Moore (OSU) and Fred Swanson (USFS). Since I was fortunate to be the very first one, many other writers have undertaken residencies at the H. J. Andrews. To see a list of them, read excerpts from their writing (including more of mine), or to find out how to apply to take part yourself, see: http://springcreek.oregonstate.edu/programs.html#LTER*

Butterflies and Battle Cries

In spring of 1985, one John Hay turned up on the roster for a butterfly-watching tour I was to lead in England. When I discovered, that first morning in London, that he was *the* John Hay, I was somewhat terrified. Long a devotee of *The Great Beach, In Defense of Nature,* and *The Run,* I stood in awe of the great nature writer. But John, a man of humility and wit, quickly put me at my ease both in field and pub.

One day near trip's end, we sought the celestial Adonis blue in a North Downs grassland beside the ancient Pilgrim's Way. June rain pelted the chalky turf. We found enormous Roman snails there, but the blues remained hidden beneath their broadleaf brollies in the meadow. Afterward, as we were drying out in a country inn over a pint of good Kentish ale, John broke the usual repartee to remark on the strange sense of disjunction he felt, practicing this most benignant activity of butterfly watching over the same fields and downs that have soaked up the blood of thousands of years of wars.

As I came to know John better, I understood why this man of keen intellect and sensibility and a deeply peaceable nature would find the discordant images of butterflies and battlefields troubling. He has hardly been alone in this emotion. From World War I came an indelible image stirred from the same insoluble ingredients. It is the final scene in the 1930 film adaptation of Erich Maria Remarque's novel *All Quiet on the Western Front,* where a soldier sees a butterfly alight on the rim of his trench, reaches out to touch it—and is shot. Dietrich Bonhoeffer, in his *Letters and Papers from Prison,* wrote of Peter Bamm: "He dreamt in a nightmare that a bomb might come and destroy everything, and the first thing that occurred to him was what a pity it would be for the butterflies." And who, of my generation, the one most marked by Vietnam, can forget Joni Mitchell's more hopeful reverie from Woodstock: "I dreamed I saw the bombers riding shotgun in the sky, turning into butterflies above our nation."

A powerful version of this profane dichotomy was given me by a late neighbor and veteran named Roy Palmer. A soldier involved in the liberation of China near the end of World War II, Roy and his company came upon an uncovered mass grave. "It looked like there was confetti falling all over it," Roy told me. "All that color!" Then, coming closer, with his handkerchief over his nose, he made out the astonishing scene: hundreds of putrescent corpses covered by thousands of butterflies, rising into the air and falling back down to blanket the bodies and to sip the sweet liquor of death, the leaking body fluids, rich in amino acids.

Far away in time and space, a precocious young student who accompanied me on many butterfly outings surprised his family and friends by opting for the Marines over college. Tyler took part in the initial invasion of Iraq as driver for the first woman to lead American troops into battle. He came home safely, but his buddy, who volunteered to stay on to clean up mines, was killed almost immediately. "Sometimes we wish Tyler had never traded his butterfly net for a machine gun," wrote Tyler's parents in their Christmas card to us.

Tyler was redeployed to Fallujah on his twenty-first birthday. Since then, he has been sending me j-pegs of butterflies—a few emigrant painted ladies and a little zebra blue—alighting on a solitary eucalyptus in the barren compound. He wanted to catch some for a better look, so he acquired a net from an army mosquito specialist and fashioned a pole handle for it. Recently, he sent me a photo of himself in camo fatigues, holding up his butterfly net in one hand and his assault rifle in the other, with a big smile. I have all faith that Tyler will come home safely, and that the butterflies will have helped him to find his own peace in the midst of crazy war.

A few days ago, I had some time to spare at Washington National airport, so I walked outdoors to take some air before my flight. Late, low sun cast a golden light through the summer haze laced with airliner exhaust. The garden court consisted of a promenade between a clipped lawn and a row of steel trellises mounted by crowns of wisteria foliage. There wasn't a blossom to be seen. Yet as I walked down the row, a black swallowtail flew out of the foliage and circled overhead, threatening to drift over the terminal onto the perilous jetway. But it kept returning to the safe, if sterile, courtyard and dropping onto the lawn, where I glassed its sharp yellow and orange spots and little galaxy of blue scales against the night-black field of its wings. When it again took wing and sailed over me, I saw the undercarriage of long legs hanging down beneath the delta wings, just as when I'd first been entranced by a black swallowtail gliding over a Colorado ragweed patch nearly fifty years before.

Apart from the sheer grace of its presence, the most compelling part of the vision was this: at the end of the terrace, one could gaze over the entire Mall, from the Capitol to the White House to the Vietnam Memorial to the Washington Monument. As the swallowtail soared back and forth, the full panorama spread out beneath its wings. I reflected on what these icons meant to me. Long ago, all those white marble mansions and memorials were a source of pride. Some of them still are: the Library of Congress, the Smithsonian, the Jeffersonian and Lincolnian confections. But forty years ago, I grew confused, as the ideals of our republic gave way to bellicose madness. Now, these grand white buildings are once more blurred in a miasma of war. So there it was again, drawn in a swallowtail's wing-beats: that sharp scarp between warfare and the most pacific of visions—like the red-spotted purple I once watched drinking and basking at the Mall's reflecting pool, hard by the long black wall of fifty-eight thousand Americans dead in Vietnam.

When at last I boarded the plane and settled into my seat, I plugged in a headset, turned up the R&B channel, and closed my eyes. A mental massage by Johnnie Winter's guitar preceded a feature on singer Lionel Richie, who in an interview said this: "Each generation chooses to war its way through instead of enjoying our time together." Of course, the generations overlap, some more warlike than others; and it is often a case of an older generation sending the younger to do its violence. (As my Kentucky grandfather said, "If them as started wars had to *fight* the wars, there wouldn't be no wars.")

But the basic truth of Richie's lament hung in the air like the sour tang of lingering jet fuel. I took out the final sheaf of a manuscript I'd been thrilled (and still a little terrified) to read for comment: a new book by John Hay titled *Mind the Gap*. Right off, I came across a classic Hay passage, beginning with a splendid meditation on gannets, moving into an homage to Edward A. Armstrong's *Bird Display and Behavior*, and culminating with what he took for Armstrong's "deep concern for what the war did to destroy a sense of humanity in our world."

As I drifted off, this curious sequence of events played in my mind against the backdrop of our current war, and against the larger reality that the present rulership, holding power without benefit of election, wars its way through everything: the butterflies, the land, the people, the world itself. I wondered, can we find a time when the sight of a butterfly need not harshly clash with our national symbols? When a young man with a butterfly net can hunt hope in a mountain meadow instead of a dusty foreign redoubt behind razorwire and armor? When our suicidal obsession with force yields to cooperation, and arrogance finally gives way? Our very sense of humanity depends on it.

November/December 2004

NOTE: *Tyler came home safely, and is now a wildlife biologist.*

The Moth Blitz

The mercury vapor lamp cast an eerie glow in the gentle Tennessee night. Behind the bulb hung a white sheet adorned with a dozen rosy maple moths—two-inch French-vanilla-and-strawberry silkmoths that clung to the cotton with pink furry legs. Then a really big moth spiraled in, striking lepidopterist Dale Schweitzer mid-back. "Well, I know it's not a regal," said Dale, "by the smell." Lacking the pungent, nutty scent of an adult hickory horned devil, or regal moth, the intruder was instead an imperial moth—as big as a small bird, its wings daffodil yellow spattered with heliotrope. I'd wanted to see these special moths of the eastern deciduous forest ever since I was a child poring over their portraits in the books of W. J. Holland and Gene Stratton Porter. But to Dale, the gathering array of brown, gray, and beige moths, many no larger than the imperial's brushy antennae, commanded far greater interest. For we were taking part in the third Great Smoky Mountains National Park Lepidoptera Quest—the "moth blitz"—where variety is much more important than looks.

Biologists rue "the taxonomic impediment"—how little of the world's diversity we really know. How can we conserve organisms that we can't even identify? Yet molecular and biochemical science is eclipsing traditional systematics, the study of relationship among whole organisms. Enter the bio-blitz, where naturalists gather to document as many species as they can in a given area over a specified period. The first blitz was conducted by Smithsonian biologists right there in Washington, D.C. Discover Life in America amplified the idea in the Great Smokies park through its All Taxa Biodiversity Inventory, intended to inform park managers while encouraging the next generation of taxonomists. University of Connecticut professor David Wagner pioneered blitzes focused on moths plus butterflies, which are essentially fancy moths that fly by day. Dave appointed me leader of the butterfly TWIG (taxonomic working group). So while most blitzers went mothing, I got to spend forty-eight hours in mid-July chasing butterflies around the half-million-acre park.

More manic at their peak than any Christmas bird count or Fourth of July butterfly count, bio-blitzes nonetheless retain something of a gaming aspect. As the countdown got underway at Sugarlands Visitor Center, I was reminded of a military campaign, with Dave Wagner as field marshal. He dispatched some forty pros, students, and volunteers to different fronts (habitats and sectors), doled out matériel (mercury vapor lamps and white sheets, ultraviolet light-traps, nets, GPS devices), and reviewed logistics for dealing with tourist traffic, bears, copperheads, and rangers who might demand permits in an environment where insect nets are normally anathema. Excitement mounted as the three o'clock starting-hour approached. "Lisa Vice, run a sheet in Cosby Creek Campground," Dave barked. 'We really need people over at Purchase!"

The teams deployed in the widest variety of habitats possible—the Smithsonian's Michael Pogue even packed batteries and lights to high-elevation sites on llama-back—and prepared for a long night's vigil. The moth-ers' work, like their quarry, was mostly nocturnal, but as butterfly leader, mine was a day job. With my volunteers, I headed to Cades Cove, where old fields are coming back as meadows full of native plants. The bright litter on a damp, sandy road resolved into mud-puddling blues, sachem skippers, sleepy oranges, commas, swallowtails, and many more. Our task was to net two specimens (designated vouchers) of each species. By sunset we'd found some twenty species, and I joined Dale's moth group to see the action at the bedsheets.

When the morning came, Lisa, Steve, Andrea, and I set out for Newfound Gap, a high point with a dynamite view over miles of Cold Mountain country. Andrea netted a purple-glazed skipper, our sole *Panoquina ocola*; then Lisa nabbed a day-flying pyralid moth, black-dotted yellow and as big as a clothes moth, which I tucked away in a vial. Who knew whether anyone would find another? At Towstring Horse Camp, a knot of little Halloween-colored butterflies avidly sucked on horse apples. Called harvesters because their larvae feed on woolly aphids, they are always a good find. And in a nectar garden planted by Job Corps workers on the North Carolina side of the park, we chased bright buckeyes and variegated and gulf fritillaries to swell our count. When word got out how much fun we had, Dave accused me of trying to steal all his volunteers.

The Moth Blitz

Chasing silver-spotted skippers through flowery meadows may be more appealing than sorting dead moths, but in fact an elegant assembly line was underway at Sugarlands. First, each trapful of specimens receives archival labels keyed to its site. Then experienced volunteers sort middens of moths into major taxonomic groups—noctuoids, arctiids, geometers, bombicoids, gelechioids, butterflies, and so on—a table for each. Now the Opti-visored specialists go to work making hundreds of specific determinations. One labeled voucher is cryogenically preserved in liquid nitrogen for later reference. The other one revolves to a pair of Canadians taking single legs for DNA analysis and then to the digital imaging table. Each specimen finally comes to rest in the park's or some other approved collection, as an absolute baseline of what's here, right now.

On the last morning, our butterfly team made a final foray up the Little River. We sought red-spotted purples and Diana fritillaries, both mimics of the pipevine swallowtail. No such luck, but we did watch seventy-five iridescent black-and-blue pipevines sipping from a river-splashed rock—a stilling vision. I would love just to watch or catch-and-release the insects. But positive identification and DNA analysis require specimens. Most insect populations are immune to direct take, so prolific is their reproduction. Besides, the moth blitz was bested nightly by bug-zappers on campground RVs and bright-light gas stations in Gatlinburg, while every motorist's windshield exceeded our butterfly catch. Not to mention the sad, shiny pavement of swallowtails crushed by the constant parade of SUVs.

Back at Sugarlands we found that a mother on a sun-break had sighted the missing red-spotted purple, and a graduate student had snagged a male Diana. At the press conference Dave called for summaries and discoveries for each category. The final count reached 788 species, 42 of which were butterflies—about the same twenty-to-one proportion you'll find all over the world. Both figures set new bio-blitz records, with several moths new to the count, and the olive hairstreak added to the park's impressive list of 115 butterflies. The result of all this bustle: the largest taxonomic sample ever documented by DNA. The task now is to extend the inventory to other parks, other realms: to census our neighborhood of life as if it really matters.

Later, as overtaxed taxonomists dispersed for beer and pizza and teachers combed the leftovers for classroom specimens, I found the little vial in my pocket and handed it to Don Davis of the Smithsonian, who took a bleary-eyed look. He and Dave Wagner raised their collective eyebrows over the unfamiliar mite and concurred that in all likelihood it was species number 789, and perhaps even new to science. Some twenty-five new species of moths, more than five hundred species of all groups, have been discovered since the Great Smokies Bio-Blitz began. One tiny moth might not mean much in the big picture, but that's how diversity evolved, and shall be known, or lost: species by species.

January/February 2005

Consolation Prize

On November 3 a front was on its way out, leaving the rivers high and the sky clear. On a saddle between the Naselle River and Salmon Creek, yellow sun struck the shellacked leaves of salal and a late cluster of its flowers, hanging like little sugar-dipped bells. I got out and breathed deeply, then drove on. One more watershed gap brought me down to Deep River, where the Pioneer Lutheran Church shone in fresh paint and local cedar shingles for the first time in many years. I followed a steep and shadowed forest lane to Deep River Cemetery. Bright light leaked out of a recent clearcut beyond the edge of firs and hemlocks, glancing off frowsy bracken and bright fake flowers. The third growth clearcut was nothing unusual in this timber-dependent district, but the sun it admitted was a grace note where November usually comes in heavy grays. Mossy tombstones of those who felled the big trees long ago, and lichened monuments of the farmers who followed, ran in loose rows up toward the shaggy fringe of forest.

My eyes ran over the eroded stones. There were lots of Wirkkalas, Rangilas, Piriilas, and Takalos in this Finn-rich ground. A plain gray pillar bespoke "Gustaf Gustaffson / Apr 16, 1890- Feb. 10, 1910." Just twenty years. A low stone with a time-sheared lamb told of "Infant Daughter, Infant Son, Infant Daughter of John E. and Christen Rull / Died Nov. 3, 1886 Dec. 21, 1889 Dec. 19, 1891." And nearby, "Kristine Rull Died Jan 12, 1895 Age 14." Little wooden markers stood for Austin, Elrin, Bill, Mary, John, and one inscribed just "Rull." A walk through an old graveyard full of children too soon taken always puts one's own depression in perspective. And yet, it says, we live, for now.

Apart from the people, the cemetery was full of life. The dignity of the spruces, the irreverence of nuthatches beeping and nutting high up among spruce cones the color of their own breasts in the sun. Two ravens roaming the air all around, shifting their shadows from the sanctified turf to the down-to-earth slash. Two bald eagles pumping overhead toward an uncut knoll. A big darner dragonfly, tardy migrant, glistening back and forth

119

through the clearing. Countless shimmering strands of ballooning spiderlets, their sinuous sway on the suggestion of a breeze mirrored by a long waving sword fern left solitary on the cutover slope beyond. A squirrel ululating from a hemlock wall, and a lively otter carved in red granite. Sprays of artificial autumn leaves reflecting the Crayola motley of real vine maples above. As I headed homeward, my postelection blues ran down Deep River in a pink wash of dogwood, cascara, and cherry. The appalling results would remain, but so would I, better able to face them for the tonic of the hills.

There are days when things look so bleak that I'd gladly trade in the whole world for spare change. For me, such occasions included the death of my mother at fifty-two, the shooting of Bobby Kennedy and John Lennon, the grounding of the *Exxon Valdez,* and the presidential elections of 2000 and now, 2004. Times like these, I instinctively go outdoors.

Not that a ramble afield makes everything okay, just like that. But going out, alert and open, causes some chamber of the heart that has temporarily drained to pump again. You remember that you can harbor loss, hold tight to sorrow, and honor grief, while still rejoicing in the rich gifts of the Earth. I think this is what Bruce Springsteen had in mind when he said in the plenty-dark song "Badlands" that "it ain't no sin to be glad you're alive." In a life with its fair share of darkness, I have found full-body baptism in the plain and glorious particulars of life to be a powerful antidote to despair. The fact is that the details of our natural surrounds offer infallible fascination and a route out of morosity. In a world deeply flawed by the infantile excesses of our own kind, this is no small potatoes.

So it was one recent morning, a bright day darkened by reports of faraway war. Returning from the mail walk, I tried to corner an unseasonably singing bird in a blackberry copse. It concealed itself among curious towhees and song sparrows, and when I tried to pish it up, I was quickly surrounded by a couple dozen kinglets, chickadees, and Steller's jays flocking into the plums overhead. Then a soft burr gave away a Douglas squirrel twitching nervously, upside-down on a huge black walnut trunk. Its belly the same butterscotch as the low beeches below, the squirrel hopped all over the tree, shot along an arcing limb to oaks across the road, and raced down to the ground. As the chickaree disappeared into the brush, a female northern harrier materialized over the field below, quartering silently for voles; and up behind the house, a bright white ornament atop an evergreen resolved into a white-shouldered kite, basking in the sun. As I came nearer, the kite dropped toward the river, its wings bent into a silver sickle. When, reluctantly, I finally went inside, the day felt safe from anything.

There used to be a quarterly supermarket magazine called *Ideals*, whose thick, glossy pages depicted scenes of perfection: frost on the pumpkin, turkey on the table, tulips by a farmhouse porch. Lush photographs and anodyne verse created a state alloyed from equal parts Currier and Ives, Norman Rockwell, Hallmark, and Ranger Rick—everyone's ideal of a rural realm shorn of rough edges. Such compositions do exist in the actual world, and I watch for them. I saw one the other day, beneath the same black walnut where the squirrel disported. A neighbor had passed below the house on the valley road, taking her granddaughter and her small dog for a walk in the warm afternoon sun. The low-angled light shone through their long, bouncing blonde hair as they walked, through the dog's woolly white fur, and through each yellow leaf of the towering tree, setting them aglow. For just that moment, all was well in the valley; or so it looked to me.

When I was a boy, my grandmother loved *Ideals*. I too was drawn to its balm for the common longing. I remember the sting of youthful abashment when I learned that people more sophisticated than I considered such depictions to be kitschy, sentimental bromides. Maybe such a judgment is appropriate, if each "ideal" edits out what we most need to know in order to engage with the agents of change and loss. And maybe that was why I loved them. Pretty pictures, though, cannot stanch a society hemorrhaging from superstition, greed, aggression, oppression, and indifference. It's not far from idealized reality to willful blindering, as about half the voters have shown again and again, voting against their interest.

Other critics dismiss the validity of a heartful response to simple beauty as merely spiritual, sentimental, or transcendental. So be it. The nomenclature of numinous experience is of no concern to me—especially since I find my numina in phenomena, and can't even tell the two apart. Sticks and stones aside, a transfixing vision actually can redeem the world, if only for the moment. And sometimes a moment is all we need.

March/April 2005

Cosmic Convergence

Elbows resting on the wooden railing of a back-beach boardwalk, I scanned the dark swamp and its bright dapple of waterlilies. Suddenly my eyes came to rest on a composition so perfect that, were it submitted for a magazine cover, no editor would believe it was candid. There, smack in the center of a pond lily, sat a western chorus frog. As its botanical name suggests, *Nuphar polysepalum* has numerous buttercup-yellow sepals that embrace a circlet of red stamens, which in turn ring the stigma. The overall effect is a crenellated corolla, almost ruffled. The "ruffles" just circled the little frog's bottom and legs as it sat on the broad knob of the stigma: a fancy throne made especially for this particular prince. Another frog, bright spring-green like the first, commanded a dark lily pad below. The whole lemon-lime tableau lasted just long enough for me to share it with a disgruntled old hippie who came shambling along the trail. It made the day for both of us. Then the frogs leapt into the marsh.

Back home, I looked up the pond lily in Pojar and MacKinnon's *Plants of the Pacific Northwest Coast* to make sure I had its identity right. That's when I read something that first amazed me, then caused me to nod with recognition. There it is again, I thought. Convergent evolution.

When the pond lilies ripen, they spill their seeds into the water encased in a gelatinous mass—much as the frogs themselves do. The physics of this arrangement involve hydrostatic pressure, passage of electrolytes, and maintenance of temperature, moisture, and chemical regimes all favorable for the eggs, or the seeds, as the case may be. Ecologically, the eggs-in-Jello solution protects the all-important zygotes from predators that don't fancy a slimeball for breakfast. And depending on whether this fertile lugie is fixed or floating, it ensures either placement in the proper habitat or colonial mobility around the marsh.

These mucilaginous packages of protein also serve to confound and fascinate onlookers. Once, canoeing near home in a narrowing waterway called Seal Slough, I came upon what strongly resembled a decomposing human brain floating in the black water. I suspected amphibious sex rather than foul play, but it was years before I discovered that our red-legged frogs commonly put out egg masses as big as small watermelons.

This natural history is interesting in itself, but there is more to the story than a world gone curiouser and curiouser. The reproductive tactics employed by both elements of that sublime pairing, frog and lily, illustrate a principle of natural selection with stunning examples all around us. Essentially, when something works well in nature, it tends to arise more than once; in some cases, many times. Moths, for instance, have evolved ears independently at least seven times in response to the echolocation of bats—some species have tympanic membranes on their heads, others on the thorax, still other groups on the abdomen. Another amazing case of convergence, involving birds acting like mammals, pounced on me from the Sunday comics. I've been reading Mark Trail's simple natural history homilies for some fifty years, and I still learn from them. In one particular strip, Mark talked about "pigeons' milk," a nutritious fluid that both pigeon parents regurgitate for their nestlings. Richer than cow or human milk, this liquid furnishes critical protein for the squabs. Pigeons' milk is manufactured in the lining of their crops beneath their breasts, but I can picture a future refinement, bypassing the bill altogether, as virtual mammae erupt from beneath the feathers. Pigeon nipples? Why not? Stranger things have evolved.

Frog phalluses, for one. Most amphibians have undistinguished cloacas for genitalia. But the tailed frog, a denizen of rapid waters, has acquired a mock-penis to ensure that his genetic gift won't be washed away. Confronted with such anomalies, systematists (scientists who try to unravel the true relationships among organisms) must decide whether two life forms with similar characteristics are monophyletic (have a common ancestor) or polyphyletic (have come to resemble each other via separate evolutionary paths). To do so, they seek to distinguish between features that are homologous, having evolved from the same root structure, and those that are analogous, or unrelated and merely similar. The wing of a bat and the hand of a human are homologous, as is obvious to all but creationists.

But the aedeagus of a swallowtail, the penis sheath of a tailed frog, and the phallus of a chimpanzee—all evolved for introduction of sperm to the female, and all equally effective—are elegant analogies. As the writer Brad Leithauser described convergent evolution in his remarkable verse-novel *Darlington's Fall*, "The eye, for instance—look how Nature kept / Contriving it anew, freshly seeing its way / Out of the darkness."

Thus organisms arise that act or appear much alike, even though their relationship may actually be remote. One must look way back in evolutionary time for common ancestors between frogs and waterlilies. In fact, recent DNA analysis suggests that animals may be closer to fungi than to the higher plants. But consider another wetland pair, much more closely related yet still disparate: otters and newts. Walking around the shore of a small lake near Olympia one spring, I had the uncommon experience of watching both animals at the same time, the newts in shallow water by the shore at my feet, the otter farther out. I was struck by their shared shape, motion, and function as lithe aquatic hunters, clumsy out of water. And I recalled the great otter-writer Gavin Maxwell's anecdote of walking an otter in London, related in *Ring of Bright Water*, in which one astonished passerby asked him if the animal were some sort of newt. Indeed, one can easily imagine an Aristotelian taxonomist placing otters and salamanders as kindred creatures based on looks alone.

So form follows function follows form … but which comes first? Neither, as it turns out. In *The Machinery of Nature,* arch-ecologist Paul Ehrlich says that convergence reveals "how evolution tends to produce similar organisms where conditions are similar," because "the morphology (form) and physiology respond evolutionarily to the characteristics of their physical environment." So what comes first is the land, the water, the world in which these organisms dwell and adapt. Since the world is infinitely variable, it is not entirely surprising that we have mammals like birds (bats, duck-billed platypus), like fish (whales and dolphins), even like reptiles (pangolins and armadillos). Convergence runs broad, and deep.

But how does all this concern us, apart from a certain Ripleyesque ripple of gee-whiz thrill offered up by a weird world? Quite a lot, in fact. We have converged on nature in our technology time and again; snowshoes arose when some early Beringian tried to chase snowshoe hares in deep powder, and Velcro took its cue from certain seed stickers. Yet the relevance goes deeper than mere innovative emulation, or biomimicry, as it's called these days: we share both analogies and homologies with much of life. Reading Robin Wall Kimmerer's *Gathering Moss*, I was thunderstruck to realize the similarities between the spore capsule of a moss and the human womb. Though separate since a time beyond imagining, they nonetheless sprang from a common source in our shared era of origins.

The most dangerous idea in the world is that humans exist separate from the rest of nature. The greatest enormities against the Earth stem from such delusions, just as us-and-them thinking justifies our inhumanity toward one another. But as poet Robinson Jeffers wrote in "The Answer," "the greatest beauty is / Organic wholeness ... Love that, not man / Apart from that." Our chief, maybe only, hope lies in accepting this truth, and in cleaving to, not from, that living whole.

May/June 2005

The High Price of Getting Hip

Forsaking the out-of-doors for any shopping mall is my idea of the Seventh Circle of Hell. But not long ago, on a perfectly good Montana Saturday, I found myself driven by necessity into a Target store. Having found the needful item tucked among acres of shiny products, I made my small purchase and sought a quick escape. The little skillet needed no packaging, so I turned down the proffered bag.

"Why don't you want a plastic bag?" inquired the tall youth at the checkout stand.

"Too many at home," I said, and should have left it there. But I just had to add, "Besides, way too much plastic ends up in the landfills. It's all oil, you know."

The checker returned a stare as blank as a big-box wall. In turn, I felt both sanctimonious and alien. But he thought it over for a moment, and then said, "Oh, don't worry. We'll always have enough of whatever we need." I left Target that day wondering whether I envied that callow lad's complacency.

The catalogue of clichés and aphorisms pertaining to the curse of knowledge or its absence is long and deep. On the upside, we have "Ignorance is bliss" of course, "What you don't know won't hurt you," and even "Ignorance is Strength" from the Ministry of Truth in Orwell's *1984*. "A little knowledge is a dangerous thing" voices the same sentiment from the inverse view, as does Aldo Leopold's undeniable dictum that "the penalty of an ecological education is to live alone in a world of wounds." But my personal favorite is Bob Seeger's brilliant and enigmatic lyric from "Against the Wind": "Wish I didn't know now what I didn't know then."

Sometimes ignorance is merely the state of learning-in-progress. This was the case for a long time with carbon-induced climate change. But there comes a point when the absence of understanding becomes willful. We have a whole other set of sayings for this state, having to do with ostriches and fools in paradise. Then there is the kind of ignorance that is innocent, if not altogether innocuous. Most Americans who notice butterflies at all tend to think all big bright ones are monarchs. Usually they are referring to yellow-and-black tiger swallowtails, rather than orange, tailless monarchs. This common error doesn't disturb me unduly—at least these folks are seeing, contrary to Nabokov's plaint about "how little the ordinary person notices butterflies." But the common confusion of two very different butterflies, one abundant, one at risk, does impair efforts to raise conservation awareness on behalf of monarchs.

Enter Scotts LawnService, a national company connected to Ortho and Smith & Hawken. Like ChemLawn (now part of the TruGreen empire)—another entity high on my list of should-be-endangered species—this company seduces naïve householders into a deal with the pesticide devil through the peculiarly American peer pressure arising from lawn-envy. By signing contracts for periodic lawn care, homeowners assure delivery of toxic chemicals directly to their children's play-place. Nowhere in Scotts's promotional come-on does the word "pesticide" appear, yet the company employs an arsenal of dangerous chemicals including 2,4-D, carcinogenic glyphosate, and various potent insecticides and fungicides. (Some products they used for years and insisted were safe have since been withdrawn by the EPA.) Not only does Scotts exploit our national state of naïveté over household toxics, but it does so under the lulling logo of a butterfly—one that manifests the company's intentional "ignorance" as none other could. Every Scotts van—I saw one on the way to Target—bears the clear image of a swallowtail's shape, but overlaid with the coloration and pattern of a monarch! The friendly lawn care folks accomplish this bizarre metamorphosis in the utter absence of conscience, celebrating their cynicism all the way to the bank.

So who cares? The mere twisting of truth might matter little if it concerned nothing worse than a case of mistaken identity among butterflies. In fact, that logo would be hilarious if the product were benign. But the company's hypocrisy is stunning: in its promo brochure, the "before" picture shows a diverse lawn that butterflies might actually visit, while the "after" shot depicts a sterile sward barren of butterflies or any other native pollinators.

Without invitation or warrant, company employees scan lawns and leave check-lists of "undesirable" plants present on homeowners' doorsteps--thus piquing guilt and gaining credulous customers. The plants checked off, of course, are the ones of value to pollinators: violets, clover, plantain, and so on. Meanwhile, evidence for the deleterious effects of biocides in the ecosystem accumulates faster than mercury in Great Lakes graylings. When it comes to poisoning our fish, our communities, our groundwater, even our children in the very places they should be safest of all—their own backyards!—then false advertising matters more than we can possibly know.

What sort of society tolerates such a practice? The same society in which a high-school worker in a chain store can assert that we will always have as much as we need or desire, in contradiction of biology, good sense, and abundant evidence. But it wasn't the clerk's sanguine outlook that most disturbed me that day at Target; long live his lucky, silly bliss. No, it was another, younger boy I encountered in Electronics who really got to me. On my way to the checkout, I'd paused to eyeball the kids gone to ground among the video games on that bright Saturday. I looked over one lad's shoulder to see what sort of foe he happened to be killing. To my abashment, his game of choice featured mountain biking through a range of virtual habitats—mountains, streams, meadows—all realistically portrayed. Well, I thought, how about that? He's still indoors, but at least he is experiencing the outdoor world vicariously. Maybe the game will serve to lure him afield after all.

Then I looked more carefully. The speed of the bike was picking up, the joy-stuck rider flying down the trail at a perilous and thrilling clip, dodging rocks, smashing trees and bushes, hurtling over dips and rises. The kid looked mesmerized by speed and action. I could not believe what I saw next, so I watched as it happened over and over. There were also pedestrians on the trail—youngsters backpacking, old folks strolling, joggers—heading in both directions. Each time the mountain biker approached one of these targets, he swerved to strike it. The walkers went down with splats and stars, and lay still and contorted as points racked up. Then I got it—the child was cast as killer after all: a super-cyclist offing hikers and oldsters instead of aliens or gangsters. Mad Max on two wheels.

Target left me shell-shocked, and now, thinking back, wistful. I wish that boy could go for a hike, try hunting for his own food, learn what killing means before he reaches recruitment age. I wish that blissful bagger could see the landfills, roam the rangeland given over to oil rigs, breathe the air in a neighborhood next to a plastics plant. I wish the CEO of Scotts could watch a real swallowtail laying eggs in an unsprayed yard where healthy children play. I wish I didn't know now what I didn't know then.

July/August 2005

Unauthorized Entry

After winding a hundred miles of curvaceous roads between wheat and lentil fields, traversing a sea of cereal where once a rich and rolling short-grass prairie grew, I pulled into Pullman with minutes to spare. Noam Chomsky, the revered and reviled linguist and radical truth-teller from MIT, was to speak at Washington State University on "Imminent Crises: Responsibilities and Opportunities." I was in the area, so I'd motored south to meet a friend in the Palouse Prairie town and to catch Chomsky. "Don't worry about getting here early," Bryan had advised. "The Beasley Coliseum is huge, and there hasn't been much notice. There should be plenty of seats."

I made my way to the arena, only to find the doors all closed and knots of people held at bay by alligator gates and uniformed officials. The crowd had reached capacity—or so said the fire marshal. Yet the arena could seat around eight thousand people. I joined one of the smaller congregations of the disappointed. "Are there really eight thousand people in there?" I asked, impressed by the so-called cow town's response. The guard replied that the twenty-five hundred people within filled all the space allotted for the event. High black curtains blocked off thousands of "unavailable" seats. One woman challenged the security guard, a florid man in his thirties, to contact the chancellor. But he said that word had already come down from the president's office: no more space was to be opened, and that was that. Yet this was no underground visit on the speaker's part. Professor Chomsky had been invited to campus as the university's official and distinguished guest, to give the 43rd Annual Potter Memorial Lecture. If the president and regents couldn't prevent Chomsky's speech, it seemed they could sure as hell restrict his audience.

Tensions grew as clusters of the excluded grew restive. Word came round that a large group at one of the other gates had become rowdy, attracting the police. Our own small and diverse cadre—students, faculty, townsfolk, others like me who had journeyed from afar for the chance to hear a hero—was so far peaceful. Several of us tried to reason with the gatekeeper, who had been joined by a small elderly woman wearing a security emblem who tried to console us by saying she wished she could hear the talk herself. We asked what difference we few would make, and how the fire code could possibly apply with so many unfilled seats.

"How do you think Noam would like this?" someone shouted, and then another called out, "Here comes the Man!" as two police officers strode our way. Shades of the '60s, I thought, wondering if heads were to be broken at a Noam Chomsky lecture. The young cops never went for their billy clubs, but they did order us to clear the building, before heading back toward the unruly crowd. Then the woman guard muttered something about checking on things inside and disappeared behind one of the arena doors. I noticed that she failed to pull it shut behind her, leaving it open a crack.

"So who is this guy?" asked the guard. "Why are you all so eager to hear him?" I told him a little about Noam Chomsky: that many consider him one of the few people willing to tell the truth about U.S. policy abroad, that his talk at Gonzaga University the night before had reportedly focused on U.S. plans for world domination. "So he really is a radical, eh?" he asked.

"You could say that," I replied.

"And he's pretty old," one young woman said, near tears. "We'll probably never have another chance to hear him."

"So what will he say?" asked the guard.

"I'm glad you want to know," I said, "but it's really frustrating standing here talking about what Noam Chomsky might say, while he's right in there saying it."

I paused, eyeing that tiny gap between the steel doors. "In fact, it's too frustrating," I said. "Sorry." Then I walked past the flustered guard, not sorry at all, hoping the others would follow my lead.

"Sir, come back here!" he shouted, running after me. "You can't go in! The doors are all locked!" That was the last thing I heard before I disappeared into the cavernous black. Hands out, I felt the preposterous curtain and pulled it close around me, in case the guard or the police were on my tail. Standing there in the dark, I thought of other times I'd sneaked into places I wasn't wanted. As boys, my brother Tom and I routinely evaded the ditchrider to swim away hot days in the cool High Line Canal on our eastern Colorado plains. A few years later, too poor to buy tickets, we discovered several ways into Red Rocks Amphitheatre in the foothills west of Denver—over the rocks, through the cactus, or (best of all) right through the gate when the usher went to take a pee. It didn't work for the Beatles, when the hills were crawling with cops, but time after time we managed to hear Joan Baez, Judy Collins, Peter, Paul and Mary, and other folk singers protesting the gathering war in Vietnam. Not many years later I joined the anti-war demonstrators in Seattle, righteous young fools rushing into the streets and crossing police lines that were meant to keep us out. Some paid for it with cracked skulls.

In more recent years I have sometimes blindered myself to NO ENTRY signs posted at the edge of private timber holdings in order to document butterflies and their food plants in the wake of industrial forestry. I may well commit such acts again. An island where I once found a large colony of a now-endangered species has been closed to further survey by the plutocrat currently holding title. A pirate-landing for a quick look-see in season would not abridge my scruples. This is not to say that I am insensible to landowners' property rights; I usually seek their permission, or keep out. But times arise when we ought to be forgiven our trespasses, when Woody Guthrie's oft-omitted lines from "This Land Is Your Land" make profound sense: "As I went walking, I saw a sign there; / And on the sign there, it said, NO TRESPASSING. / But on the other side, it didn't say nothing. / That side was made for you and me." I believe that barging in to hear Noam Chomsky was one such time, since his (of all people's) right of free speech, and my right to hear him, were both being compromised. I also like to think my decision resonated with his message.

When I came in, Chomsky was hilariously skewering the advertising industry. But he soon shifted to his major theme, "the deficit of democracy" under which we live. "America is the freest country of all, by many measures," he said, "but few take advantage of it." Through a devastating litany of wars, distortions, and erosions of civil liberties, he showed how our freedoms are going, going, along with our national self-respect. We are selling a certain version of freedom abroad, while our own liberty is diminishing by the day. Chomsky challenged each of us to consider these chilling facts, and then to act on our freedoms.

Thoreau, in "Civil Disobedience," wrote, "Action from principle ... changes things ... it is essentially revolutionary," words that moved both Gandhi and King. For my own small insurrection, I didn't spend a night in jail, as Thoreau did in protest of slavery and the Mexican War. Nor were there any truncheons or tear gas this time around. But Chomsky left me thinking that maybe we'd all better be dusting off our protest skills. Besides gate-crashing, I've taken to carrying placards in peace marches and picket lines again. In these belligerent and benighted times, the reasons to resist are all around, and only growing. As we said in the '60s, if not now, when? If not here, where?

September/October 2005

Losers Keepers

If my testicles were not firmly tied on with gooseflesh, I would have lost my deep male voice long before it ever cracked. I am a loser of the first degree.

It's always been this way. When my grandmother Grace took my older brother, Tom, and me to early Disney films at the Denver Theater—*Snow White, Cinderella, Peter Pan*—we walked from her house near Congress Park to the #12 bus stop and waited beside a big mulberry with whose sticky little fruits Tom and I filled our faces. After the "streetcar" brought us back—a bus by then, but to Grammy it was always the streetcar—we walked home in the shady alley behind Detroit Street, tired and way too hot to wear the jackets we were obliged to take on every outing in case one of the famous Denver thunderstorms came up. Good thing skin is waterproof, because invariably Gram had to call the Denver Tramway Company to retrieve the jacket I'd left behind on the bus.

Not long later, my car-less mother and I managed through her tactical skills to reach a close-in Front Range canyon, famous for its butterflies, via varied public transportation. Once in Deer Creek Canyon, a true Valhalla cut off from my home by many miles of the still-small Mile High City, I went wild over amethyst Colorado hairstreaks on Gambel oaks, silverspots and skippers thronging September-yellow rabbitbrush. But when we transferred buses in front of the state capitol's golden dome, my box of specimens—my best trove yet as a young lepidopterist—did not. The Denver Tramway dispatcher got used to us.

You could cover the heads of a small nation with the hats I've scattered about the globe. The pebble-gray cap left in a hot college classroom. The purple-heather deerstalker ditched during a summer deluge on a London double-decker near Big Ben on the way to see snakeshead fritillary flowers in a Wiltshire wood. Any number of straw and Panama chapeaus, and one ridiculous Hoss Cartwright ten-gallon cowboy hat that badly needed losing. And one I truly hated to forfeit: my XXX Beaver Stetson Smokey hat, acquired for and well broken in during a summer as a ranger-naturalist in Sequoia National Park. It lived an unofficial life through the end of the '60s, that era of enhanced appreciation for outrageous headdress, then abandoned me at a roadhouse in Toad River, Yukon Territory, along the Alaska Highway. In the years that followed, I refined losership into an art. The summer of '76, visiting Estes Park, I managed to misplace my glasses, camera, binoculars, and ultimately the driver's door of my Volkswagen bus, all in the same week.

But not all of my losses have been for keeps. I have been lucky beyond any sensible expectation in the return of my offerings, and not just the thirty pounds along the waistline that I've lost and found time after time. The Black Watch flannel shirt deserted on a Delta airplane in Arkansas and recovered through the zeal of a humane ticket agent who made it her mission to restore the stray garment. The wallet I've carried for twenty-five years, a patchwork of thin leather and duct tape, that liberated itself from the top of our car on a California freeway only to reach me a month later, cards truck-busted but contents otherwise intact. The object that I've lost and found the most, and most fear losing for good, is the set of fine, small Leitz binoculars given me by my former wife, Sally Hughes, when we were living in graduate school penury. They came with us from New Haven to New Guinea, and have accompanied me everywhere I've been ever since. They are as much a part of my body as the units with which I began this column, but they are not attached. I have left them on trains, planes, and boats, at security in JFK, and atop a mountain in the Sangre de Cristo Wilderness Area in Colorado just before a storm. And once in Costa Rica, would-be thieves, seeing me near tears over my missing binoculars, returned them to me, pretending they had found them.

The first of many cases that have carried these field glasses was a neat leather bag with Indian-head nickels for clasps. I once recovered that bag through a want ad in a local paper, but finally parted ways with it at a rest stop on the Oregon coast. The binos were around my neck that day, praise be! These beloved optics—which once got me into a Van Morrison concert, inspired an impromptu haiku by Gary Snyder in Okinawa, and granted me intimate visions of an infinitude of lives and landforms—have (knock, knock, knocking on wood) always come back to me. And so, I dearly hope, it ever shall be. Amen.

Even the gray, marbled Pelikan fountain pen with which I am writing these words has taken its leave and returned over and over, like the cat that came back, or a slow paddle ball. Once it disappeared into a recess of Thea's truck for six months. Another time I lost it on the *Coast Starlight* approaching Redding at three a.m., the Amtrak conductor and I on our hands and knees among the legs of slumbering passengers, looking all around with a flashlight. He found it just before I had to detrain. And again just last week: somehow misplaced at a conference in Eugene, almost given up for gone, it arrived in the post yesterday from a friend who had used it to sign a book and later found it in her purse.

Every time such beloved items find their way back home, I feel a disproportionate thrill of joy and redemption, like the simple glee expressed in the words of that old pop song: "Reunited, and it feels so good." But don't misread me. I am not a very materialistic person; my only car just passed a third of a million miles, my only boat has buoyed my bulk in its eighteen feet of wood and fiberglass for decades. Except for the beckon of books, I'm a rotten consumer. Yet I am a materialist, insofar as modest physical objects matter to me. As a biologist, I try to imagine the adaptive value for such behavior—why do we feel such affection for old belongings? Is it merely sentiment, or resistance to change, or is it something atavistic? After all, our pets and their wild ancestors have their favored lairs and playthings. The violet-green swallows return to the same nest-hole in our porch year after year. Captive chimps, like children, give up their accustomed blankets only under violent protest. Possession seems to have deep evolutionary roots. Artifacts have always mattered to people. Sure, they are only things, and when lost, they reassimilate into the world. But they are the things that recall our histories, and root us to our material existence.

The fact of the matter is we are all losers, even if not as practiced as I. We lose our special childhood places to subdivisions and shopping centers. Our hard-won freedoms, along with the most basic expectations of a civilized society—education, healthcare, and security in old age—are slipping away from us, filched by officials and legislators in service to wealth and power. In large ways and small, we share the certainty of loss, every day and all through our lives—our elders, our hair, our keys, all gone. Yet the losses we suffer deepen our gratitude for the good things still in our keeping, and make us cherish those that come back from the brink: recovered binos, the ivory-billed woodpecker, your lover rescued from the grip of mortal illness.

We keep nothing, of course, beyond our temporary tenure. And if possession is only a short-term loan, then what we call loss might be seen as an act of early return, like taking a library book back before the due date. I'll remind myself of that the next time I have to buy new specs, having taken to leaving a trail of trifocals all across the land.

November/December 2005

NOTE: *Many more recent examples appear in my book* Mariposa Road: The First Butterfly Big Year, *including one spectacular recovery told in a chapter entitled "Lost and Found."*

Squirrel Tales

It was Harrison Cady, not Thornton Burgess, who put the topcoat on Old Mr. Toad, the trousers on Buster Bear, and the crooked stovepipe hat on Sammy Jay. I'd always thought the author of the famous *Bedtime Stories* and Mother West Wind tales had imagined his animal characters thus clothed. But an exhibition of Cady's original drawings at the Cape Cod Museum of Natural History set me straight. Cady thought Burgess's animals would be more appealing to children if dressed in human accouterments. Apparently it worked; the *Bedtime Stories* became enormously popular during the first half of the twentieth century.

Thornton W. Burgess (1874-1965) had a special genius for creating characters—Paddy the Beaver, Jimmy Skunk, Bob White, Little Joe Otter, and the rest—who spoke and acted like people while behaving true to their animal natures, in their rightful habitats: the Green Meadow, the Smiling Pool, the Briar Patch, the Laughing Brook. More children discovered natural history through Burgess's wonderful stories and Cady's enchanting drawings than in any number of field guides or textbooks. I was one of these. Between the ages of five and ten I wore out every Burgess book in the Aurora, Colorado, public library. My very first title, the only one I owned myself, was *Chatterer the Red Squirrel,* published in 1915. So began a lifetime's love affair with squirrels.

Over the years, from Chatterer to the western gray squirrel I watched on the way to a train this morning, I've learned that not everyone holds squirrels in equal affection. To many, a squirrel is a squirrel and by any other name is still just a squirrel. Some consider them "rats with bushy tails." And it's true, they can get into your attic and make a racket and a shambles. They are also capable of ravaging birdfeeders worse than jays, though their brilliant acrobatics to defeat "squirrel-proof" feeders are surely worth the price of a bag or two of the best black sunflower seeds. But sentiments toward squirrels vary as much as the animals themselves. Classified as everything from game animals to furbearers to varmints, invasive pests to endangered species, North American rodents of the family Sciuridae are a multitude, in Walt Whitman's sense of the word.

From Chatterer, I graduated to watching big, brushy eastern fox squirrels in my grandmother's Denver garden. Grammy fed them peanuts, and we looked on enthralled from her breakfast nook as they nabbed the nuts and scatted up to the high wires of telephone lines, their els over the city. Because our raw young suburb was still almost treeless, we had no squirrels at home in Aurora. So I ranted till I received a much-coveted, life-sized plastic model of an eastern gray squirrel. You had to sprinkle a fuzzy powder over a skim of airplane glue to create its pelage. I wasn't very good at models—my B-32 and USS *Forrestal* were a mess, and the squirrel came out worse. It's a wonder, having made that fur, that I didn't end up a glue sniffer. Still, I loved my scabrous squirrel. Along with Chip 'n' Dale comics, it furnished my sciurid fix between visits to Gram's house.

I did not yet know that those Denver yard-squirrels were no more native to Colorado than we were. The arrival of eastern gray and fox squirrels in western cities followed white settlement, irrigation, and the planting of eastern hardwoods as ornamental and shade trees. Their invasion traces the classic pattern of an adaptive species happy to live alongside humans, opportunistically following the progress of civilization across the landscape. The eastern gray (*Sciurus carolinensis*) and the bigger, rustier fox (*S. niger*) seldom share territory, but divvy up the turf like rival gangs. Frank Richardson, my University of Washington mammalogy professor, discerned for example that eastern grays occupied the university's main campus, while foxes held forth in the arboretum across Seattle's Montlake Ship Canal. A melanic population of eastern grays thrives in Vancouver, BC. Fox squirrels throng Missoula, Montana, as common as skateboarders. In honor of their city's abundant eastern grays, the good people of Longview, Washington, have erected both a giant wooden statue of a squirrel clutching an acorn and a squirrel overpass above a busy street, modeled after the Tacoma Narrows Bridge and dubbed the "Nutty Narrows."

Most conservationists believe that these successful colonists have, like the white people who brought them, displaced the natives. Chatterer was a red squirrel (*Tamiasciurus hudsonicus*), first cousin to the western Douglas squirrel (*T. douglasii*), common among Douglas-firs. Both species are often called chickarees for their piercing scolds and churrs. Generally, when eastern gray and fox squirrels show up, chickarees split. In Britain, the lovely russet, tassel-eared European red squirrel (*T. vulgaris*) has likewise dropped out as the coarser American grays have come in, like a reciprocal gift for starlings.

It turns out, however, just as many Burgess stories do, that the actual plot may not be what it seems. Some ecologists now question whether the alien squirrels are really responsible for driving out the reds. People commonly assume that European cabbage butterflies have excluded native mustard and checkered whites. But as Tufts biologist Francie Chew has shown, the cabbage white adapted superbly to anthropogenic conditions, while native white butterflies found such human-mediated habitats no longer suitable for survival. Such an effect may apply to squirrels as well: eastern gray and fox squirrels followed the folks, while chickarees fled with the retreating wild. Thus the wildly adaptive squirrels have become a conspicuous symbol for our own modifications to nature.

Along with Chatterer, the Burgess cast includes a gray squirrel, too. Happy Jack is bigger and brasher than Chatterer, but equally likable. Burgess's readers, many of whom were concentrated in the populous American northeast, would have been familiar with both: Chatterer the chickaree in the piney woods, and Happy Jack the gray in the farm country and towns. But over time rural conditions retreated, and the balance of power shifted. As development grew beyond all bounds, the descendents of Happy Jack out-adapted Chatterer's kind. Perhaps anticipating such changes, Burgess founded the Green Meadow Club for land conservation and the Bedtime Stories Club for wildlife protection, and late in life, he received a special gold medal from the Boston Museum of Science for "leading children down the path to the wide wonderful world of the outdoors." What higher honor? Still, the landscape changed, and *S. carolinensis* became the generic American squirrel that most people picture today. As with squirrels, so it goes among our own species: we trade Main Street for Wal-Mart, the greasy spoon and truck stop for Wendy's and Shari's, the mom and pop for the Kum & Go. Today, homogeneity rules among both strip malls and squirrels.

Yet regional squirrel species do retain their strongholds, even now, in the wilder precincts. In the Rockies live the splendid charcoal Abert's squirrels with their great fluffy ear tufts. Along the north rim of the Grand Canyon, the striking Kaibab squirrel haunts the pinewoods, flashing its white tail over its black pelage. I'll never forget the abundance of western gray squirrels (*S. griseus*) I encountered daily as a ranger-naturalist in Sequoia National Park. This biggest and showiest American squirrel, with its ashen pelt and huge silvery plume, stands more than a foot tall when reaching for an acorn; but the western gray is also the slowest, and I grieved at the number of flat ones plastered to Sierra roads. This species still thrives in California and Oregon, even in towns if they have plenty of oaks; but it dwindles to a seldom-seen gray ghost up in Washington. Species usually thin out near the edges of their ranges—edges that we erode daily.

I actually live in something of a squirrel nirvana. Our pioneer farmstead and its old hardwoods have been found by both eastern gray and fox squirrels--one of each. Appalled, my wildlife manager friends urge me to shoot them immediately. But unless squirrels have learned to bud, these singletons represent little threat. They will pass; and in the meantime, along with Townsend's chipmunks, flying squirrels, and native chickarees, they make of our place the most squirrel-diverse spot in the state. I shall enjoy them while they last.

In the meantime, happily, here amid the hemlocks of home, our native chickarees continue to hunch in their classic squirrel poses, tails curled over their backs like pelted punctuation marks, gnawing spruce cones and black walnuts. I often hear them chattering, and sometimes, when the light is right, I think I can see their little red vests.

January/February 2006

NOTE: *The gray and the fox have, indeed, passed from our precincts, most likely recycled as redtail, raven, or great horned owl.*

Hanky Panky:
Notes on the Biology of Boogers

There's nothing like a sinus infection to concentrate one's attention on the head ... and how nice it would be to have it removed. Same goes for a common cold, or first-class hay fever. The itchy eyes, water-faucet schnoz, chapped nostrils, postnasal mudflow, mucous membranes swelling and shutting off the breath like a clothespin to the nose—the sheer discomfort of it all is enough to make you Google "guillotine."

The worst thing about this mucosal madness is the utter fecundity of snot. *Where does it all come from?* I contemplated this question throughout my childhood. As a boy I was prone to periods of snorting and hawking that would have outdone any old snoose-shooter at his spittoon. My folks wondered aloud why their son was so phlegmy, and my grandparents muttered old-fashioned words like "catarrh" and "rhinitis." Once my third grade teacher, Mrs. Haner, reprimanded me in class for noisy snorting, and I was so embarrassed that I refused to return to class the next day. The issue escalated into a standoff between myself, armed with my pogo stick atop our swing-set slide, and my imploring mother, followed by a friendly but firm policeman, and finally a grim truant officer who looked like Broderick Crawford and almost scared me snotless.

In summer, I suffered torturously from hay fever. This was diabolical, since I was always out of doors in search of butterflies, which abounded in the pollen patches. Sometimes my eyes swelled nearly shut from all the rubbing, and my throat itched so badly that I found relief only by grinding Sugar Pops with the back of my tongue. I'd sneeze thirty or forty times in a row. In winter, strep throats and tonsillitis took turns until a tonsillectomy dislodged the wen of bacteria gone to ground in my ravaged glands. When I left home after high school, I quickly realized that much of my misery had been due to eighteen years of secondhand smoke, to which I was allergic. Severe pollution sometimes brought the same result. A London inversion followed up by country pollen once landed me in the Royal Infirmary of Edinburgh with pneumonia, and bronchitis from a memorable Mexico City smog actually induced me to write a will in my journal.

So where *does* it all come from? Mucus generates from membranes called mucosa—thin tissues that line and lubricate many bodily passages including the airways, sinuses, and lungs, the throat, stomach, and gut, the vagina, urethra, and foreskin. Mucus is secreted to protect surfaces by trapping foreign matter—dust, smoke, pollen grains, invading organisms—and then showing it the way out. Composed of water, mucin (large glycoproteins), inorganic salts, epithelial cells, leukocytes, and granular matter, mucus also facilitates functions from swallowing to sex. Both saliva and tears contain it, and crying bunches the membranes and increases the flow. When we are well, and nasal effluent, or phlegm, remains viscous and thin, we hardly notice it, merely blowing our noses and clearing our throats now and then. But phlegm runneth over during illness, bouncing germs and eliminating waste from immune-system struggles, and thickening with pus when inflammation draws out an unholy gumbo of used white blood cells and septic fluids. (Can the consistency of okra be a mere coincidence?) When infection or histamines engorge nasal mucosa, the familiar soccer-ball-in-the-snot-locker feeling results. Decongestants may shrink mucous membranes, but they leave behind a raw Saharan burn and do little to promote actual healing.

It is perhaps some consolation that our species is hardly the only one blessed or plagued by phlegm. Sinuses arrive in the fossil record with crocodilians, and all the vertebrates since have possessed them, so they aren't simply the jape of a petulant creator, as it sometimes seems to me. In fact, nasal or otherwise, mucus is an evolutionary marvel. Certain male amphibians and mammals protect sperm and ensure paternity by installing a mucus plug in their mates' vaginas. But the most sophisticated mucus users are the terrestrial gastropods we call snails and slugs, which secrete slime trails from their broad "feet." The slick stuff allows them to traverse gravel, or even a razor blade, literally greasing the skids. I sometimes wonder if the ox loggers who basted their corduroy skid roads with bear grease got the idea from watching the banana slugs (*Ariolimax columbiana*) so abundant in the forested hills they labored amongst. Slug and snail lubricant serves as propellant too, since it possesses the amazing ability to be whale-oil slippery when secreted, then instantly switch to super-glue sticky, so the muscular mollusk slides and glides, slips and grips, covering ground with greater celerity than one might expect.

Mucus also figures both in slug sex and predator avoidance. The extra-slimy European brown slugs (*Arion ater*) that bedevil West Coast gardens mate, sometimes *en trois*, with a gob of congealed mucus surrounding their protruded and mingled genitals. And just as a throatful of the stuff can mortify the cold-sufferer, a would-be predator can gag on a slug in full ooze. Thea once discovered a garter snake in our woods unable either to swallow or spit out a slime-swollen *A. ater.* When Northwest Indians roasted banana slugs escargot-style, they removed the mucous glands first. But *A. columbiana*'s heroic level of slime production has also rendered it especially useful to humans. As the late Ingrith Deyrup-Olsen of the University of Washington discovered, the copious mucus production of banana slugs makes them ideal partners for research on cystic fibrosis, a genetic disease whose sufferers labor under a surfeit of thick, sticky mucus secreted by overactive cells.

Wonderful adaptations notwithstanding, slugs both native and alien often elicit revulsion and cruel treatment. Perhaps this is understandable in light of our attitudes toward our own bodily secretions, which we so determinedly seek to mask or abolish, employing a vast array of sprays, swabs, capsules, suppositories, and other nostrums. Few people delight in boogers, except young boys enamored of their utility in arousing exhortations of *Ewww!* or *Gross!* Yet we would all perish from desiccation, starvation, or infection in no time, not to mention abstinence and consequent extinction, absent our personal slimes. And when our own lubricity fails us, we invest in still more pills, ointments, and jellies.

I take heart in the happy knowledge that my hay fever has diminished with age, making for much better butterfly summers. And that Washington State has just activated a ban on smoking in all public places. I celebrated with a half pint of Hale's Wee Heavy Ale in the Blue Moon, a famous Seattle pub rich in Beat, '60s, Dead, and Grunge associations, where I danced my socks off in college but which has been off-limits for decades with its solid wall of smoke.

Best yet, I have discovered an elegant and low-tech solution to a superfluity of snot—a simple little ceramic vessel shaped like a genie's lamp and commonly used in India for nasal cleansing. Called a neti pot, this ancient yogic implement allows one easily to irrigate the sinuses with warm, salty water. This is not nearly as unpleasant as it sounds, and it confers sinus health, easier breathing, and cold protection at least as effectively as our usual arsenal of vitamins, herbals, and other placebos. Twelve bucks, no drugs, fewer lugies: now, that's my kind of medicine. It lets me go lighter on the hankies, too. And though I know I'll always be a Man of Mucus, maybe when I croak they won't have to use the epitaph, "He fought the good fight, but the snot got him in the end."

March/April 2006

Feathered Remnants

The first bird I saw upon awakening in Central Asia was a common myna. Perched on a high wall, it contributed its brown, white, and olive to the roses' coral and the blushing orange of persimmons in the courtyard of the Atlas Guesthouse. Several more of the yellow-billed, yellow-legged mynas, like robust thrushes with cowlicks, joined the first. Over the next several days, I seldom climbed up or down the wrought-iron spiral staircase to my room without one or more mynas hanging about, or slicing open the Asian air. Like black kites over the Ganges, egrets on the industrial fill of San Francisco Bay, or kestrels haunting the verges of interstates, motorways, and autobahns, these mynas bespoke a long indwelling with people in their millions, turning the dense human overlay to their advantage. They became a signature of the handsome, post-Soviet city of Dushanbe.

One of the measures by which I've always parsed the days and places of life has been the birds that have accompanied me, not just as ticks on a life list, but as companions in the business of being somewhere--in this case, Tajikistan. I'd come here with Laurie Lane-Zucker, director of the Triad Institute, to teach a ten-day seminar on place-based writing for the Aga Khan Humanities Project. My writers were twenty-one university students and professors from cities in Tajikistan, Kyrgyzstan, and Kazakhstan. "Stan" means "land" in all these Turkic languages, and we had come together to discuss the land itself, our place and lifeways upon it, and how writing can help conserve it in the twilight of traditional farming and in the glare of fierce global change.

We worked in both English and Russian, and sometimes Tajik, Kyrgyz, and Kazakh, employing gifted and hard-working translators. Communication was painstaking, deliberate, and exciting. An exercise could take all afternoon. When we attempted transliteration of the last paragraphs of *Lolita*, these writers made a literary, moral, and esthetic leap entirely new in their experience. And when I played Nabokov's poem "The Swift" for them in the author's own voice, in both Russian and English, they were beside themselves with delight. Following his example, I asked my writers to attend to the birds, because the ways they adapt to the land can signify much about our own potential for adaptation, because of their inherent beauty and interest, and because the colorful mynas were all around us, everywhere we looked.

Returning to the guesthouse after my first day in Dushanbe, I found the evening sky full of crows. Carrion crows, crossing the sky back and forth to who knows where, enlivened the smoky dusk with their dark shimmer and coarse exhortations. Among the "caws," I heard another corvid song that I well remembered from my years in England: the falsetto, two-part call and response of the onomatopoeically named jackdaw. "Jack!" one would call, and "Daw" answered another. Scores of these compact black birds with their dapper, dove-gray heads spun among the clouds of crows. Venus hung above the treetops, blazing as I'd never seen her before, with a star just above. Suddenly, as I was glassing the planet and her companion, one of the jackdaws cut right between Venus and the star: a gift from the sky such as you'll never see unless you watch birds. You'd miss it too if you fell for the common trap of despising crows for their racket, ubiquity, and cheek. Their success is worth attending to: there is a reason they say that England will last as long as there are ravens at the Tower. Corvids persist in spite of all we dish out.

For organisms lacking the ecological chutzpah of crows and mynas, survival is dodgy on the tattered margins of heavy human occupation. On my only previous visit to Central Asia, to Turkmenistan (then the Turkmenian S.S.R.) in 1978, I took a boat trip on the Kara Kum Canal. Curtains of reeds blocked the view of the endless desert beyond. Even where camels trampled the banks to drink at the murky water it was hard to see beyond the mountains of cotton stacked for transport, sometimes a hundred feet high. Yet there were also chartreuse-and-blue bee-eaters cruising back and forth across the bow, a brilliant and endlessly engaging procession from reed to reed. It was good to know that the bee-eaters had survived the Kremlin-imposed cotton monoculture thus far. But nowadays, long after the Soviet Union packed it in, this vast, arid region still faces serious loss of land and water to enormous irrigation projects for cotton. Uzbek cotton culture has progressed to the point of virtually dewatering the Aral Sea itself. How long can the bee-eaters, and the great gerbil, the caracal cat, and the other endemic creatures of the Central Asian desert and steppe persist under this incarnation of King Cotton?

The participants in our workshop faced massive environmental challenges back home, from severely polluting gold mines in Kyrgyzstan to extensive nuclear test and waste sites in Kazakhstan, all visited upon these former satellites by the Soviet Union before its break-up. Yet both of their regional literary traditions, Muslim and Soviet, functioned largely in the abstract, and lacked a ready rhetoric for close and concrete personal engagement with the elements of the land and the threats it faces. Hence the Aga Khan Trust's hope to stimulate a more direct language of encounter, better suited for advancing the cause of conservation and restoration.

Our workshops were immensely gratifying, but also grueling. Between subtle cultural slights and apologies often involving speeches, and the tedious pace and strain of translation (sometimes with hilarious results), everyone felt both exhilarated and wrung out. By the time we broke for field trips, we really needed them.

Our first foray took us out of Dushanbe past a big, dusty concrete plant, up into the fringes of the tall and snowy Varzob Mountains. I had often dreamt of the alpine butterflies of the high Pamirs on the Tajik-China borders, but the season was too late for that, so Varzob stood in. Sycamores ran up the canyon in as many colors as maples, greens to reds with every tone of brown and yellow in between. Elaborate dachas occupied the lower reaches, retreats once for Soviet apparatchiks and now for the elite under the current regime. The head of the ravine opened out into a small farm with fruit trees and a goatherd's mud hut that might have come from a thousand years ago. The shaggy, curvy-horned graziers, in all the hues of the dust and rocks they foraged among, roamed every inch of the surrounding mountains, valleys, and ridges. As if forever, along with donkeys and cows, the goats had hammered and shat the alpine turf into a soft medium devoid in late fall of any plants but grass-straw and thorny weeds.

Golden eaglets flashed their white windows overhead, redstarts their russet wings and rumps among the rocks. At the base of a great boulder a rodent had left scat and the shells of walnuts and pistachios. From a split in the rock like a crack in the middle of Asia, I spied on brightly striped rock buntings, rosy chaffinches, and hawfinches—large, uncommon fringillids feathered in warm grays and browns with big white bills like those of related evening grosbeaks. I'd seen only one before, forty years ago, in Germany. Now here were dozens, plainly arrayed on the ground beneath hawthorns, feeding on the yellow fruit called haws that gave them their name. In spite of the goats, come spring, these bare slopes would burst forth with wild tulips and hyacinths. At least these few specialized birds and plants have persisted throughout the ages of grazing that has both battered and fertilized these ancient lands.

My students were having a fine time too, dispersed as far into the high hills as their young legs and the hours would take them, snapping photos off toward the towering Pamirs and Tien Shan while playing with one another and my writing assignment. Their various languages and cultures mingled in laughter over the shaggy white goat-dog that had adopted a clot of them. In heels and skirts, denim and trainers, they all bounced and whirled with the pooch. This was not their usual academic experience.

Our other field trip took us to the ancient fort and medreseh at Hissar. This was an even drier, more hammered-by-time landscape. Yet it was lively. Local tourists scrambled up among the ruins of the fort and respectfully pored over the Islamic school, now a museum. Children filled water bottles from a small but holy spring, downstream from a herd of miniature sheep browsing the little marsh. Colorful wedding parties blending dress and music of East and West took pictures before the great memorial, an obligatory visit for newlyweds. We all remarked thick skeins of ballooning spider silk that filled the air above the fort, as shiny in the sun as the flowing wedding gowns. And when someone joked in English that this spiderweb sky must be the real reason that this older-than-dirt route is known as the Silk Road, everyone—Tajik, Kyrgyz, Kazakh, "White Russian," Muslim and Orthodox alike—got it, and laughed together.

After the workshop, Laurie and I decompressed for a couple of days in Istanbul. Clouds of rock doves swirled around the Hagia Sophia, same as at San Marco in Venice, St. Paul's in London, St. Patrick's in New York. We won't eradicate birds. There will always be species able to survive, even flourish under the heaviest human impacts. But we should never

149

be satisfied with mere remnants of our avifauna, or anything else, as we carry on flattening the world. Mynas are as doggedly adapted to that weary countryside as the persimmons, the goats, and the people themselves. I enjoyed them; but I took greater pleasure in the hawfinches, still foraging the wild haws of Varzob.

In the evening, the sky above the Blue Mosque was criss-crossed by a hundred white gulls soaring on thermals rising from the massive dome. Among a forest of minarets, at the highest point, a single jackdaw perched between the horns of the golden crescent topping the mosque. "Jack!" it called. "Daw," replied another. I took it for a prayer, or plea, that if the dust ever settles on the human experiment, there may be something left beyond the lowest global denominator.

New

Pele and Kamehameha Dance

When I first visited Hawai'i in 1979, I was working for the Nature Conservancy, managing preserves in the Pacific Northwest. Since the organization did not have a land steward on the islands, I was sent to investigate a proposed preserve. In order to protect the widest possible range of biological diversity, the plan was to incorporate habitats *mauka-makai*—from the mountaintops to the sea.

The windward coast of the Big Island was largely devoted to sugarcane in those days, and the upper slopes of Mauna Kea were close-cropped pastureland thick with cattle and English skylarks. But some excellent *koa* and other native forest survived in between, and the hope was to rebuild the top and bottom reaches around the intact middle. The forest remnants were rife, however, with feral pigs, cattle, and horses, and draped with a pink passionflower known as banana poka. These Andean vines with their yellow fruits hung so heavy as to break down the very trees, while the pigs vigorously grubbed out the rootstocks of tree ferns, crucial to the forest structure for the soil they hold and species they support. In a typical Hawaiian dilemma, the biological legacy was overwhelmed by adventitious exotic species, shoved into pockets, and threatened by aggressive interlopers. I was supposed to come up with some viable management options— no picnic, since the vine seemed indomitable and both native and *haole* hunting groups wanted the pigs to stay.

More than a third of the indigenous avifauna had already been lost, but it was still possible to find some native birds by prowling fragments of original habitat. I saw the Hawaiian hawk (*'i'o*) and the Hawaiian thrush (*'oma'a*) in relatively undisturbed forest in the heart of the proposed preserve, east of Mauna Kea. Above Kilauea Caldera, on the slopes of Mauna Loa, lots of scarlet *'apapane* and one equally red *i'iwi*, with its impossibly long, curved yellow bill, probed the crimson bottlebrushes of *'o'hia* trees for nectar.

I also hoped to see the Hawaiian butterflies—both of them. Yes, these most isolated islands in the world have given rise to only two endemic species of butterflies, both exceedingly beautiful. Blackburn's bluet, or the Hawaiian blue (*Udara blackburnii*), is steely blue above, emerald green below. The Kamehameha butterfly (*Vanessa tameamea*), a relative of the much smaller painted lady, displays vivid vermilion among the black and white spots of its wings. The bluet was named for an important early explorer of the islands' insect fauna; the *Vanessa* for King Kamehameha the Great (circa 1758-1819), depicted in statues and portraits as clad in a blindingly bright cloak of yellow feathers plucked from the wing tufts of the extinct Hawaiian 'o 'o. Though its color scheme differs, the butterfly is as grand and dazzling as the king's royal vestments. Island entomologists suggested that I seek the indigenous butterflies at Kipukapua'ulu, also known as Bird Park, in Hawai'i Volcanoes National Park.

A *kipuka* is an isolated patch of older vegetation among more recent lava flows, an ecological refugium from the destruction and a reservoir of species for repopulating other areas. I entered a clearing that held the last individuals of *Hibiscadelphus giffardianus* growing in the wild. This hibiscus relative had been saved from extinction in nurseries and transplanted here by botanists. All nine of them were badly gnawed by rats and bleeding sap. There, feasting on the leaking lifeblood of the dying trees, were fifty or sixty Kamehamehas, right at eye level or above. The big, brilliant nymphs made a firestorm of pink-orange-red as they pumped their wings to drink the sweet sap. The effect was heightened, unfortunately but brilliantly, by an almost solid backdrop of *Crocosmia* and nasturtiums, both harmful non-natives, both intense orange in their exotic flare. Nearby, the stingless nettle trees called *mamaki*, on whose leaves the larvae of the Kamehamehas feed, were in bloom. All over their small white flowers flashed and nectared Hawaiian blues. My wish to see both butterflies was richly gratified, if a little clouded by the Kamehameha spectacle coming at another species' expense.

Back on the Big Island recently for a vacation, Thea and I revisited Kipukapua'ulu in hopes of seeing the butterflies again. On the way, we watched a pair of *nene* (Hawaiian geese) stride the links of Volcano Golf Course. Captive breeding and subsequent rerelease has brought this handsome endemic back from fewer than seventy-five surviving birds. Trails and scat marked the rough where mongoose, brought in for rat control, had been hunting *nene*. As we entered the *kipuka*, introduced long-tailed blues and monarchs flitted and glided, but there was no sign at first of the natives. Predictably, the *Hibiscadelphus* grove was long gone. I feared that was it for the entire species, and wondered, were the butterflies gone too? But farther into the reserve we found a clearing with a bench surrounded by *koa* and *mamaki*, our quarries' host plants. *'Apapane* whirred from *'o'hia* to *'o'hia* overhead, and wrenlike *'elepaio* called from the tree ferns. We waited; then patience paid off. Bluets descended from the *koa* and alighted nearby, greener than the leaves. Then, visiting the yellow pea-blossoms in the top of a *mamane* tree, a huge Kamehameha spread its wings full-on in the sun.

That evening we drove down the Chain of Craters Road to its abrupt end at the jumble of a recent lava flow. Neither of us had ever seen molten rock, the very stuff of earth genesis. We'd come to hike across sharp and broken lava called *a'a* to a vantage where, after dark, flowing magma could be seen pouring into the sea. One doesn't do this lightly. Auntie Pua Kanaka'ole Kanahele, a highly respected hula master and devotee of the traditional deities, once told me to skip it if it means no more than the usual tourist spectacle. Pele, the goddess of fire, who resides in Halema'uma'u Crater, is powerful. She is best not approached without proper reverence. Besides, hot lava is dangerous: get too close, and you could be toast. We poked our way carefully, and, we hoped, with suitable respect, across the rough *a'a*. Before us, a great vapor cloud arose from the ocean where the magma, tunneled from high above, emerged at the cliff and plunged into the sea to create new land underwater. We perched on a ropy flow of *pahoehoe*, *a'a*'s smoother relative, to await the tropic dusk As dark fell, the cerise glow of molten magma ignited the night a couple of miles farther along the coast. We watched for an hour. Finally, a lava bomb showered cinnamon sparks against sea and sky.

As we made our way back with the sounds of humpback whales on one side, lava crickets on the other, Thea noticed a glow on the eastern horizon—another vapor cloud, this one over the mountain, illumined pink from below: an exhalation from Pu'u 'O'o, the source of the current eruptions. When we again mounted the road toward Kilauea, we could see both clouds, stretching *mauka-makai*. Our final glimpse of the magenta tongue licking the shore far below put me in mind of something … what was it? Then I recalled Kamehameha at the *kipuka,* and saw that same molten red all over again.

I never did find out whether my advice helped much in the care of the new TNC preserve on the Big Island. Yet revisiting Kipukapua'ulu almost thirty years later, I found the *Crocosmia* and nasturtiums (and the cabbage butterflies that feed on them) almost gone, thanks to energetic restoration efforts by the National Park Service. And maroon-flowered *Hibiscadelphus,* re-established in a new grove, showed no evidence of rat bites. To be sure, the entire Hawaiian archipelago has been ecologically ravaged for centuries, and its losses are profound. But given native cultural practices that promote diversity, the dedication of conservation biologists and managers, and the land's own resilience, maybe what's left of Hawai'i has a chance. Kamehameha remains after all, dancing with Pele across the *a'a,* as the islands are rebuilt *mauka-makai* every day.

July/August 2006

With Enemies Like These

At a recent conference on "Biophilic Design in the Built Environment," I stepped outside for an early-morning bird walk. Beyond the rustic buildings of Whispering Pines, a forest retreat in Rhode Island, warblers and orioles were singing high in the trees. The woods gave way to shoreline just as the fog was lifting, and a pair of Canada geese, four golden goslings entrained between them, motored silently across the scene.

That evening, I stood in the Minneapolis airport upon a large, round tableau of Minnesota natural history. A fall of autumn oak leaves filled one side of the mosaic; a mix of mammal tracks trod the other. In between swam a school of pike, walleye, and lake salmon, and right across the middle of the floor flew a grand Canada goose, wings outstretched, wrought of gray granite, black basalt, and white quartz. Its long neck and bill protruded from the circle like the arrow of a compass pointed south. This elegant mandala refreshed the morning's vision in my mind's eye. The wild geese on the lake and the artist's paean to goose-beauty both thrilled me, in different but complementary ways.

Leaving the mosaic behind, I turned down the corridor for Terminal F, where I was arrested at once by a bright yellow advertisement featuring yet another big Canada goose. This one stared me straight in the face, bill open and all but honking. The caption read: "Goose Problems? Flight*Control*® PLUS Repels Geese." Suddenly, biophilia had turned biophobic.

Canada geese have long bespoken wild migration, as the very emblem of heroic departure and awaited arrival, of wanderlust yet fidelity to path and place as well. But feral Canadas have super-adapted to urban habitats, trading migration for year-round residency and mutating toward well-fed lawn-ornamenthood. Which might be fine, except that they chase children and pets and poop copiously on the lawns they graze, rendering soccer fields slippery and parks less pleasing. In the process, geese have evolved in many people's minds from free soul to Flight*Control*®, from wild wonder to weeds with wings—if you define weed as any organism where it isn't wanted.

Though *Branta canadensis* is native to North America, most of our weeds are non-indigenous plants and animals, wildly successful imports lacking their natural controls. The word *weed* is most commonly applied to plants, but it pertains to green crabs as well as crab grass, to Norway rats along with Dalmatian toadflax. Ecologists and habitat managers agree that what they call "invasive" organisms represent one of the most severe threats to natural areas. What one doesn't hear much about is the positive contribution weeds sometimes render in spite of their exotic status.

This is a distinctly unpopular subject, so at the outset of this apostasy, let me state my bona fides: I am a life member of the Washington Native Plant Society, a former regional land steward of The Nature Conservancy, and I would not hesitate to push the button that would restore the North American flora and fauna to their pre-European-contact composition (though this places our immigrant selves in a distinctly dubious position). But these facts do not blind me to the ways in which exotics sometimes provide ecological function and economic services, or furnish beauty and interest in otherwise depleted habitats. For readers already shaking their heads (or their fists), compelling examples abound. European honeybees displace some native pollinators, but they also serve as generalized pollinators where specialized insects have been eliminated by poisons or other factors. Many crops depend on *Apis melliifera* for setting seed, such as blueberries, squash, and pecans. Conversely, beekeepers may depend upon non-native flowers such as the terribly troublesome star thistle, a major nectar source in overgrazed western landscapes blanketed by its stickery yellow galaxies.

Native insects too may benefit from exotic plants. Butterflies are limited in occurrence by their plant needs; if either caterpillar foodplants or adult nectar supplies are scarce, populations may plummet or drop out altogether. Though its wild habitats have diminished, the two-tailed tiger swallowtail thrives across the West today because its larvae have adapted to feed on introduced green ash and its adults to nectar on bull thistle, teasel, and even the dreaded knapweed, a.k.a. The Plant That Ate Montana. And here on the West Coast, the federally threatened Oregon silverspot subsisted largely on the nectar of a listed noxious weed, the stock-poisoning tansy ragwort, before its native meadows were restored.

Likewise, no other flower is more beloved of butterfly gardeners than butterfly bush (*Buddleia davidii*), because none attract more species and individuals than this mauve wonder brought from Asia. It is also easy to grow—too easy, often escaping and colonizing adjacent sites. In my experience, butterfly bush usually spreads to highly disturbed situations such as barren roadsides, riprap, abandoned industrial grounds, and Superfund sites, adding beauty, butterflies, and soil stability where most native plants wouldn't be caught dead. In South Wales, toxic and dangerous coal-mining slag heaps full of heavy metals proved hospitable only for *Buddleia*, on which red admirables, commas, and small tortoiseshells love to nectar; and for native nettles, the necessary host plant for all three species. And so, presto! Mordor morphed into butterfly Mecca. But now, *Buddleia* is invading native floras in the both the U.K. and the Pacific Northwest, prompting debate between alarmed land managers and defensive butterfly enthusiasts.

An even pricklier paradox in California pits conservationists directly against one another. Australian eucalypts, introduced for timber long ago and now a rampant feature of the coastal hills, are the bane of restoration ecologists. Yet West Coast monarch butterflies, deprived of their original winter shelters of redwood, live oak, and Monterey pine and cypress, have made a remarkable adaptive shift to overwintering primarily in eucalyptus trees. The California migratory monarch phenomenon might well be extinct today, absent these exotics. Euc-loving lepidopterists and native-plant purists go at it hammer and tongs over whether to retain or even plant new gum-tree groves for sustaining the monarchs until fragmented native forests can be restored to furnish the necessary winter protection. These camps often overlap, leading to cases of acute eco-schizophrenia.

Does all this mean we should abandon our campaigns against "invasives"? Certainly not: stemming the spread of exotics is some of the most urgent and important work being done today. But there is no magic button to restore the indigenous biota, and many grays lurk among the green. Our weed-fighting fortunes are sure to go up if we make rigorous distinctions. I question, for example, whether an all-out pogrom against a species with such a devoted clientele as *Buddleia* is sensible when available resources are too few for more pressing threats such as gorse, garlic mustard, Japanese knotweed, and leafy spurge. Instead, I advocate a vigorous but targeted response where an alien species gains a beachhead in native habitat, such as the Jamestown S'Klallam tribe's efforts to banish butterfly bush from gravel bars on Washington's Dungeness River.

Unlike some of my dedicated friends—and this might be a character flaw—I can appreciate individual animals and plants outside their native context: a cabbage white where no other butterflies fly, a skylark singing over Hawaiian grassland where it has no business nesting, a Scotch thistle in an American hedgerow. These interlopers help me to remember that only one species bears culpability here, the same one that must now clean up its nest after centuries of bad decisions by its littermates.

By all means let us battle the weeds—recognizing as we do that they sometimes have an upside. And let us proceed without hatred of the lifeforms involved, for they are doing nothing more than that which evolution has asked them to do, and which we humans have both allowed and abetted.

September/October 2006

NOTE: *As I'd imagined, this essay attracted some strong feelings from people who felt I went too easy on exotics. My sharpest critics were a conservation biologist and a land steward, both good friends of mine for whom I have immense respect. They made some strong points in their letter to* Orion. *I accept that I was overly sanguine in places, especially about butterfly bush. This plant has now grown to dominate a number of important riparian corridors, been listed as a noxious weed in several states, and should not be encouraged. These days I urge butterfly lovers not to plant* Buddleia, *unless it is one of the new sterile varieties; and if they already have it and insist on keeping it, to be sure to deadhead the flowers before they go to seed and to police their precincts for escapes. I stand by my basic arguments that we should take each case on its merits, and not demonize the plants and animals themselves. When I wrote this essay, I had only rarely heard mooted the potential ecological benefits of certain introduced species. Since then, many papers and editorials, entire issues of biological journals, a book or two, and several conferences have all been devoted to assessing this very question. All this palaver can only be to the good.*

The Great Indoors

On a recent airline flight, having finished one book and forgotten to bring another, I read a piece in United's *Hemispheres* magazine by Patrick Thorne, titled "Outside In." The opening spread showed a scene I assiduously avoid, a throng of human beings arrayed in attitudes of leisure upon a tropic beach. A closer look revealed clearly phony coconut palms and an overarching … not sky, but *ceiling*. "The outdoors is moving indoors at some of the world's largest and newest resorts," enthused Mr. Thorne, in the commerce-speak common to such rags, "with exciting implications for the future of vacation travel."

I was transfixed like a moth at a porch light. "Do you fancy snow-skiing during the summer? Or how about some tropical sunshine to go with the snow in winter? Or maybe you'd just like to have a rain-free vacation guaranteed? No problem."

No *problem?* The appalling truth is that such antidotes to an inconvenient world actually exist, and more are in the works. Thorne acknowledges that such facilities as indoor ice rinks and swimming pools have existed for a long time. "But some of these new resorts feature a rink that's 20 times larger atop a snow-covered slope or a pool the size of a small sea with tall waves where you can surf every hour on the hour." And that ain't all! "Throw in an elegant hotel," he oozes, "plenty of retail shopping, and fancy restaurants under that very large roof, and you'll appreciate the scope of these indoor/outdoor resorts." Well, some might. Or rather, many must—or the billions these synthetic environments cost would never be spent. After all, as the author asserts, these are places "where money can be made year-round."

Almost every day I see, hear, or read something that makes me feel as if I've somehow missed the whole damn zeitgeist—if not the evolutionary limb to which I'm supposed to belong. Hummers and Cadillac SUVs, big-screen TVs in bars and restaurants, the blogosphere in general, and nationalisms of all kinds whether played out on battlefield or soccer field—these I just don't get. But few features of modern life give me such a sense of separation from the culture at large as the hell-on-Earth hideaways that so beguile Mr. Thorne.

To be fair, I am not among their target clientele. I will take almost any day out-of-doors over almost any day within. As I write, Thea and I have just returned from North Cascades National Park and Okanogan National Forest, where the unpredictability of the weather, biting insects, rough roads, steep trails, wildfires, wind, and washouts all affected our days and nights afield. The closest we came to an elegant resort was either a five-dollar campsite at Harts Pass or the Country Cabin Motel in Quincy—it's a tossup. And apart from gasoline and fruit stands, our retail shopping consisted chiefly of buying a bandanna at the Mazama Store and a six-pack of Busch N/A "beer" as legal tonic for the long drive home across the overheated plain after a fortnight in high, cool places. We encountered fewer fellow vacationers than we would see in a single visit to Safeway, and had to settle for refreshing ourselves in lonely mountain lakes and streams instead of stewing with a crowd in tepid tropic waters. To these deprivations should be added the psychic toll of exposure to sixty species of butterflies and at least that many birds; pikas and marmots, chipmunks and squirrels, deer and fawns; and miles of roadside verges lined with tall white orchids and china-blue lupines. It was rough.

Our recent experience is further incomparable to the artificial paradises in that the pleasures of our days came largely in the form of the unexpected. Sure, there is a tremendous buzz to be had when you predict the presence of an animal based on botany and maps and your own field experience and desires, and then the beast actually shows up. But the unanticipated encounters—discoveries, if the word may be stripped of its imperial baggage—bring a joy still keener than the fulfilled quest. Seeking Freija's fritillary, find instead a giant sawfly? *Cool!* In contrast, the outside-in resorts promise an experience where "All of the uncertainties of a traditional vacation are removed. If you want snow, you get snow. You needn't worry about the elevation of your resort, a sudden thaw-inspiring warm spell, or any extreme weather"—not to mention global warming.

The Great Indoors

I suppose these artificial destinations might be considered cousins to the butterfly houses I support and enjoy; after all, both revolve around fabricated habitats and come up short on surprise. But surely there is a significant difference between Butterfly Rainforest in Gainesville, with its ecological and educational mission, and Tropical Islands, a vast, domed rainforest near Berlin, replete with a half-mile winding path and camping on a bogus beach. Or the Ocean Dome at the Phoenix Resort in Japan, "where a water screen generates a 230-foot-long, 40-foot-high sheet of water on which lights, images, and lasers are projected for maximum dazzle"; Ski Dubai, where a 1,330-foot-long, black-diamond ski run abuts the 350-store Mall of the Emirates; or the $155 million Swedish passenger liner that simulates a tropical cruise in northern waters in winter, under a protective plastic hatch—all of which already exist. Xanadu, a $1.3 billion indoor snow confection in New Jersey, and the $5 billion Dubailand, complete with a dinosaur park, are under construction.

Though I love the butterfly houses for their wafting longwings and shimmering morphos, I would swap my *n*th face-to-face with a rice paper butterfly under a glass roof for any *plein-air* encounter with a little copper or skipper in its home habitat. Curiosity flourishes only when the outcome of experience is in question, which happens through random, grounded encounters in the real world. We learn most from nature when we meet it on its own terms.

Of course "reality" these days is relative, as is a sense of what constitutes worthwhile experience. Some have argued that these fake places might serve as honeypots to reduce the pressure of the masses on real habitats, much as multifamily dwellings relieve pressure on undeveloped land—and that they could save fuel, by allowing thousands of New Yorkers, for example, to ski in New Jersey instead of Colorado (though their own operations must suck energy like water). Others might dismiss the likes of Xanadu and the Phoenix Ocean Dome as merely another decadent display of wealth beguiling a credulous and easily amused public. But I think they present a real danger, to the extent that they actually meet—and condition— people's expectations of experience. If enough folks fall for the blandishments of the bogus, will it not erode humanity's concern for the authentic? Few will vote or pay for what lies outside their own perceived needs, and when the building tab for one of these techno-wens exceeds the total budget for the national parks and forests combined, surely they can't promise anything but ill for the future of the actual landscapes that support life on Earth. Seeking a simulacrum of perfection in the ultra-ersatz will only insulate people from our imperfect yet wonder-filled world, and hasten its destruction.

On that note, Stephen Hawking recently suggested that the collapse of ecological systems may oblige us to seek new homes in space. If he is right, maybe the counterfeit resorts can furnish tips for the endeavor. But I, for one, do not believe there will ever be anywhere else for us to go, nor any reliable way to get there. Even contemplating such an out distracts us from caring for the home we have. When it comes to planets, there really is no substitute for the original.

November/December 2006

The Territory of Tint

The color gray appeals to me, or perhaps I should say the full spectrum of grays, from pearly pigeon-breast gray to ashy or granite gray to weathered cedar-plank gray. And I like it spelled that way: g r a y. Just as well, in both cases, since I live in a place called Gray's River, which was named after Captain Robert Gray but could easily take its name from the panoply of leaden, pewter, and old aluminum skies that ceil this rainy place. I delight in a cloudy, foggy, or mist-ridden morning. In fact, I take keener relief from a cool gray break in a too-long stretch of overheated, UV-saturated, blue-sky days than the reverse.

Even so, I consider the human trait of color vision to be one of the greatest gifts of kindly evolution. As a kid, I loved the fact that our license plates bore the motto COLORFUL COLORADO and that, according to our fourth-grade teacher, Mrs. Frandsen, the name Colorado referred to our state's red rocks. I was crazy for Crayola, and the bright construction paper from which we fashioned turkey tails and autumn leaves for our classroom windows. Given this childhood infatuation with the rainbow, it isn't surprising that seashells and butterflies captured my fancy, or that I asked for parrot tulip bulbs for my eighth birthday.

In "Kodachrome," Paul Simon sings, "Everything looks worse in black and white." While I don't entirely agree—penguins, pandas, and polar bears, old Hitchcock films, and December days in Gray's River would all suffer from colorization—I echo his sentiments when he sings, "Give us those nice, bright colors." Or, as Cezanne put it, "Long live those who have the love of color—true representatives of light and air!" I find no conflict between this view and my penchant for hoary hues. After all, the very author of the gray I celebrate most days, the rain, sponsors as well the richest gathering of greens you could find anywhere. And my PhotoGray glasses not only lessen the glare of the sun, they also saturate tint. Yet we couldn't even consider such questions were we not imbued with vision across the portion of the electromagnetic spectrum—lying between the ultraviolet and the infrared—that we call "color."

Some people, when they discover that color vision is not general in mammals, feel bad for their pooches and pussycats. Yet our pets know nothing beyond their limited chromaticism, and even if they did, I'm not sure they would swap the exquisite sensitivity of their smell and hearing for what they might regard as the cheap trick of a parti-colored existence. Of the subtle range of perception they achieve through their noses and ears, we know nothing more than they know of our near-deaf progress through a colorful world. We haven't even a word for nose-blindness! Yet anyone who reads Henry Williamson's classic *Tarka the Otter* or Daniel Mannix's *The Fox and the Hound* will apprehend something of our mammoth ignorance of these alternative sensory systems.

And what about the colorblind of our own species? Should we feel sorry for them? Two of the best butterfly observers I know, biologists Janet Chu and Paul Ehrlich, have orange-green colorblindness. In fact, Ehrlich's research subjects of choice have been checkerspot butterflies—orange animals of green habitats. Perhaps their color-sight "aberration" (from our viewpoint) gives such people keener pattern recognition than the fully color-sighted ordinarily enjoy. Certainly such traits can vary. My wife, Thea, and illustrated-journal artist Hannah Hinchman both possess incredibly perceptive eyesight. Thea routinely spots four-leaf clovers in full stride, and hiking with Hannah in Idaho, I was astonished at what her peepers picked out. Yet both women's acuity flags at the onset of dusk; each, in fact, has markedly poor night vision.

Our visual abilities, including the perception of color that we so often take for granted, arise through specific populations and configurations of rods and cones in the retina of the eye. Certain invertebrates, such as bees and butterflies, also see in the color spectrum, via clusters of ommatidia— parallel fiber optics that convey images as light waves through the thousands of lenses in their compound eyes to their optic nerves, and on to their brains. But these creatures tend to see in the ultraviolet as well, which we cannot. Observers often assume that a yellow crab spider secreted against a yellow daffodil achieves invisibility from its prey, but we cannot know how a UV-sighted insect sees the spider. Its camouflage has actually arisen to confer safety from its own color-sighted, avian predators. Conversely, the pink-and-purple dots and lines in the mouths of flowers, called nectar guides, fluoresce in ultraviolet, looking like neon strips to their insect pollinators. As a sharp student in one of my butterfly classes remarked, "Oh, I get it. Landing lights!" I often hear people declare that butterflies prefer this color of nectar flowers over that, but I mistrust such opinions because what looks yellow or mauve to us might look otherwise to an insect. UV perception often trumps color in mate recognition, too, in butterflies.

Among the vertebrates, we humans share color vision chiefly with birds, which also enjoy a degree of UV perception, according to current research. Color vision in humans and other old-world primates came about for adaptive reasons, often thought to include our omnivory. In his fascinating book on the evolution of human eating habits, *Why Some Like It Hot*, chile-pepper aficionado Gary Nabhan explains how peppers have coevolved with herbivores. Birds, insensitive to the fiery compounds with which hot peppers are graced and craving the carotene they contain, seek out peppers and pass the seeds through their guts intact. Mammals, whose digestive systems would harm the seeds, typically find capsaicin unpalatable and learn to avoid peppers. Humans are the exception, likely because of the power of peppers and other spices to preserve food and deter parasites. Those benefits may have outweighed the peppers' oral burn, which many people have even come to enjoy. Presumably, birds pick out pecks of peppers by recognizing their bright red colors. It is interesting that some of the only other vertebrates that can discern chile peppers by color have come to value them as condiments.

Certain aspects of our relationship to the territory of tint strike me as even more curious. For example, why the peculiar adaptation of a sense of beauty, as it relates to color? Why should we thrill to a rainbow? Swoon before a coral sea? Experience, upon viewing a scarlet tanager for the first time, what the ornithologist Arthur Cruikshank, Sr. once described on a field trip as an "ornigasm"? I cannot say. I only know that when I see the russet fletching of a sharp-shinned hawk's breast against the cones of a Sitka spruce, or a red-tail's tail flashing past last year's fronds of a winter cedar, something about the scene seems eminently right.

Nothing clashes in nature, but certain colors just look good side by side. For example, the cherry crown of a redpoll among rosehips deep in winter's bleakness on a white Wyoming plain. Spring azures nectaring on bluebells, and swallowtails on lilacs at Easter time, as Paas-bright eggs hide in the fresh green grass below. The magenta of Parry's primrose in the alpine and Lambert's locoweed in midsummer meadows. Then in September, Wilson's warblers staging like lemon drops among scarlet currants before migration, and cinnamon monarchs floating over fallen peaches. I still love a winter's monochrome on the lower Columbia, and admire the array of grays that gives rise to every rainbow. But I am no less thrilled to awake each day to a world produced in Technicolor.

January/February 2007

Evolving, Swiftly

The first rain in weeks slickened I-5 as Thea and I drove south to see one of the great spectacles of northwestern natural history: the Chapman School swifts. Every September, as they stage for their migration to Central America, Vaux's swifts congregate in the heating-plant smokestack at this 1925 Northwest Portland elementary school. When the importance of the avian bivouac was first recognized and publicized by Portland Audubon in the 1980s, the students, staff, and administrators of the school elected to wear sweaters and leave their furnace off until the swifts departed, saving thousands of birds' lives. In 2003 the heating system was replaced, but the stack left intact for the benefit of the birds and the many citizens who flock to admire them. Each evening during the staging period, folks gather on lawns around the chimney, much as the people of Austin assemble beneath the Congress Street Bridge to picnic and cheer the fly-out of a million Mexican free-tailed bats that roost there. As in Austin, a festival atmosphere reigns at Chapman, one barely dampened by the long-awaited rain on the evening we took part.

The showers also failed to discourage the swifts, or the insects on which they feed. We saw some six thousand birds working the airspace above the school (at the peak of the phenomenon, as many as thirty-five thousand birds dazzle appreciative watchers). Their dark gyre spread and spun over the neighborhood, then drew in toward the haven as light dimmed and, one by one, the birds disappeared into the stack. A Cooper's hawk alighted on the rim, causing the remaining airborne swifts to shriek and spread out. Then the hawk dropped into the black hole, snatched a sacrifice, flew off to eat it, and the swifts resumed their circling queue for fly-in. When the hawk came back for seconds, the swifts scattered once more. At last the raptor left, and the remaining swifts spiraled down inside that lightless pipe, like a plume of smoke running in reverse, swirling back into its stack. As the final bird fluttered, stalled, and then just dropped into that black throat, I thought helplessly of Tom Swift.

Tom, hero of a series of juvenile adventure books launched in 1910 by Edward L Stratemeyer, is known for his clever inventions, his zip, and his pseudonymous author Victor Appleton's love of tortured adverbs. A classic construction in a Tom Swift story might read: "'I've found the missing jewels,' Tom exclaimed brilliantly"; or "'It actually flies!' he loftily announced." Such locutions came to be known as "Tom Swifties," and considered a field mark of overwriting beginners. It was partly in reaction to these easily parodied modifiers that many editors and writing instructors adopted the conviction that, without a darned good reason for being, most adverbs must die. How could I not think of Tom Swifties, as I beheld those thousands of birds falling spectacularly, darkly, *swirlingly* into a hole in the night?

Swifts, with their easy flutter, instant takeoff and turn, rise and fall, flick and weave, romance the air like no other fliers—not swallows, not falcons, not bats. People often conflate swallows, passerine birds of the family Hirundinidae, with swifts, the Apodidae, which are more closely related to hummingbirds than to sparrows and swallows. Through parallel evolution, both groups have acquired pellet bodies and cutlass wings. Swifts, however, are longer and slimmer of body and wing, even more adept in flight, and—in a word—swifter. Swallows can function on the ground, landing and launching to gather mud with which to create their nests. Swifts, with their nearly vestigial feet (Apodidae means "without legs") cannot walk, only cling. So they spend most of their time on the wing—feeding, migrating, even mating aloft. I doubted the latter claim until one day at Joshua Tree National Park I witnessed a pair of white-throated swifts erupt from a boulder cleft in tandem, the male mounting the female in midair.

I have also watched white-throated swifts carrying bits of cellophane on high above granite domes in the Sierra Nevada, dropping them, catching, rising, dropping, again and again, in sheer play. I've gazed up at squadrons of sickle-winged black swifts migrating over high passes in the Cascades. And I've taken great pleasure in packs of the species known in England simply as *the* swift (*Apus apus,* the sole species there), as they scream past tolling bell towers and over fish 'n' chip shops at dusk.

Swifts are largely urban birds in Europe, as is the chimney swift (*Chaetura pelagica*) in eastern North America. This preference for settled sites stems from a remarkable adaptive shift in these species' habitat requirements. Swifts, sky-bound sifters of the aerial plankton, rely on cavities for roosting, nesting, and chick rearing. Inside, they cling to walls with their claws, braced by their spiny tails, and affix bracketlike nests to vertical surfaces with sticky mucous (the source of "birds' nest soup" in China). When most of the big old hollow trees vanished during the Middle Ages in Europe and later but more rapidly in colonial America, the swifts should have diminished, or even dropped out—yet they remain abundant today. Like metropolitan peregrines nesting on steel I-beams and feeding on feral rock pigeons, Old and New World swifts found a substitute for their vanishing support systems: chimneys and smokestacks. The European swift was also called "devil bird" for its sooty, stygian dwellings, and soon after the eastern American species was described, it acquired the common name "chimney swift."

Vaux's swift (*Chaetura vauxi*), a western species, abounded around tall dead trees when pioneer ornithologist J. K. Townsend described it in 1839. Both tree boles and broken branches, reamed out by rot, harbor the swifts. But as old-growth forests have shrunk around the West, capacious hollow trees have grown scarce. A western red cedar candelabra might last for centuries, but another species favored by swifts, black cottonwood, decays much faster—creating habitat quickly, but ephemerally. It is the entire mosaic of growing, dying, decaying, and regenerating trees that has served the swifts so well in the West, and it is just that mix that has been compromised by industrial forestry with its short rotations, and an associated increase in the size and intensity of forest fires. As a result of both, Vaux's swift has turned up on several state lists of species of concern in recent years.

Yet this bird too has been showing signs of an adaptive future. The first site purchased with Washington's personalized license plate habitat fund, in the 1970s, was an old square smokestack at an abandoned mill on the Klickitat River where migrating swifts roost each fall. Evidence is also beginning to accrue of Vaux's swifts actually nesting in chimneys. In 1944 Rachel Carson observed that "the western cousin of the chimney swift—Vaux's swift—only of recent years has begun to make the transition from trees to chimneys." Reported cases of chimney nesting were still uncommon half a century later when, on a rainy, fireplace kind of day in May, our neighbors Joel and Noreen heard swift chicks peeping in their chimney.

I would rather have our swifts move into town than pass from the scene altogether, but I'd far prefer to see them remain country birds that only occasionally visit cities. For as delightful as city swifts may be in Vienna, or St. Louis, or either Cambridge, they represent the failure of human cultures to maintain complex woodlands. Out here in Vaux's country, we could end up with an agreeable mix of swift habitats—school chimney and forest giant—but only if we adopt a silviculture that maintains fully functioning, recycling, ancient forests, as well as preserve chimneys and silos where healthy forests are lacking.

No doubt, it is exciting to witness adaptation in other species firsthand, but there is a dimension of this story that applies to humans as well. We too are losing vital elements of our own habitat—soil, water, and space, for starters. Will we also get a second chance? As I watched that dusky vortex settle into the Chapman School chimney, it occurred to me that we'd better start doing some evolving of our own—and swiftly.

March/April 2007

Book Tourist

When you take part in the archaic but still-kicking enterprise of making and selling the original laptops (by which I mean books), you hear certain questions again and again. What about blurbs, folks ask—are they paid for? No, *never*. And what's the deal with royalties? *What's that?* I ask back, since for most authors significant royalties are a thing of the past. But what a surprising number of readers want to know about is that mysterious phenomenon known as the book tour. As I set off on another of these odd peregrinations, the subject is much on my mind too.

Authors commonly grouse about the ordeal of the book tour—as if hanging out in treasure-houses, having folks you don't even know make a fuss over you, and sharing your own stories with willing listeners who actually pay good money for your books and want you to sign them were somehow worse than wrangling a mouse all day in some bleak cubicle or fighting in some dope's war. Methinks such scribblers doth protest too much.

Okay, maybe nobody throws their panties onstage, and any publicity trek is admittedly fatiguing and economically barren, except for the hypothetical casting of bread upon the waters. But in my experience book tours aren't that bad, and can even be fun—at least the kind I do. Mind, as a solidly "mid-range author," I have never been called on to make the twenty cities in twenty days, sandwiched between Charlie Rose and Oprah, kind of a glitz-blitz tour. Maybe those marathons really are as awful as the gripers make out. Yet taking a new book out into the world is not only a rare chance to air and share that which you've worked damned hard to create, but also to interact as a social creature in the complex ecology of bookselling after many months of solitary labor. At a long-ago awards banquet, Barry Lopez spoke of "the community of readers and writers." I've never forgotten that lovely term, and I would add librarians and booksellers to this charmed assemblage. The book tour is the ecotone where all these mutually dependent organisms commingle: the magic terrain where the habitats of scribbler, peddler, and reader meet.

It helps that I eschew the big chains. While few besides Idaho congressmen any longer characterize predators as evil, the tactics of the books-by-the-gross folks have truly been predatory in the old Red Riding Hood sense. Like introduced brown snakes on Guam or marine toads in Florida, these voracious gobblers have critically unbalanced the economics of bookselling. Diversity and stability are fostered by a plurality of publishers and shops, not by the monoculture of corporate mergers and empires. A friend who is not a mid-range author—a perennial bestseller, in fact—objected fiercely when her publisher's sales contract obligated the house to "deliver" her for one signing with a certain big-box outlet. Only a spontaneous bout of "food poisoning" saved my friend's no-chain principles. By sticking with the indies, writers get to call on shops that care about their books, not just the fees for displaying them that the chains demand.

One of my earliest tours was allotted a budget of $400. The publicist had in mind roughly three big-city bookstores and hotels. I asked if I could manage the money and the itinerary, and in the end I visited thirty-five independent Northwest bookshops between the Canadian and California borders. Many of these, at least those that have survived the onslaught of the chains, have become friendly habitats where I am always welcome and my books are kept in stock and hand-sold by merchants who know their customers' tastes—a highly desirable and truly coadaptive state of affairs for any writer.

For my current book, I will be calling on booksellers from the Skagit Flats to Sisters, from Bonners Ferry, Idaho, to Ilwaco at the mouth of the Columbia River. It will be fun, like showing a new grandchild around the 'hood. It will also be full of encounters with remarkable people who persist in the economically unrewarding yet deeply nourishing traffic of ideas, images, stories, and experience, printed on the transformed pulp of trees, bound between hard or soft covers for the benefit of hungry readers— literary ecology at work. Literal ecology too, as it turns out, because the smaller shops tend to survive, even thrive, in rural locales beyond the market grasp of the biggies. I reach them on country roads, over passes and across prairies, via ferries and high bridges. One such constellation clusters around the shores of Puget Sound and hems of the Olympic Mountains, in view of scoters and orcas, madronas glowing red at sunrise against high firs, fog-swept headlands and sun-struck strands. I call this leg my Highlands and Islands Tourlet.

As for the carbon debt that touring entails, I travel by Amtrak whenever possible. I conducted my *Chasing Monarchs* tour almost entirely by rail, and shall never forget monarchs nectaring trackside on yellow tarweed as the *Coast Starlight* swept northward from Santa Barbara. Beyond the reach of rail, I drive my small, gas-sipping car from town to town. I like to think that the energy I use is somewhat offset by placing books directly in people's hands. When readers come out and buy a book at a reading, that's one less plastic-wrapped parcel ordered on the Internet and shipped long distances at the expense of the courageous shop struggling to serve its community.

Not that it's all a blast. Each of us has our own Worst Tour Story, often involving mixed-up dates and empty halls. One rainy night in Salem, my audience consisted of two faithful friends, one casual patron, and a street person who came in to get warm and a free glass of wine. At any given time during the reading, at least one of them was sound asleep (but never all together). Another time in Grants Pass a single person came, but she, the amiable bookseller, and I spent a memorable eve discussing my book and others. In the community of readers and writers, "wherever two or more are gathered" truly applies.

The book tour is also a Cinderella story—a brief ball before being cast back into the word scullery. On publication day for *Where Bigfoot Walks*, I even had a "handler" to usher me around Seattle to several interviews and the launch at the Elliott Bay Book Company. My literary escort entertained me with stories of which famous authors were nice and which ones were just pricks, as opposed to those who couldn't keep their pricks in their pants. When I got to Denver, my thoughtful publicist, Mab, kindly put me in the Brown Palace—a grand, triangular sandstone hotel that as a child I'd dreamt of someday waking within. But when Avis wouldn't accept my maxed-out Visa, I had to hire an '82 Mercury lowrider with a sprung hood from Rent-a-Wreck on Colfax Avenue. When I rolled up beneath the canopy at the

Brown Palace, the doorman said, "Don't even worry about a tip, Bud." I had a full house at the Tattered Cover, readers as well as ringers—all my high school friends in town for our thirtieth reunion. Then Mab called to say I'd need to stay an extra night for a mandatory radio interview. "There will be three to four million listeners in thirty-six states," she said.

"What's the program?" I asked.

Mab paused for a moment, then said, "It's called *Weirdo Night*. You're on between two alien abductees." When I asked if at least I could stay on at the Palace, she said, "Nope. Budget's all gone. Pick any cut-rate motel on Colfax." Since the interview was scheduled for three a.m., I skipped the motel and just napped in my pumpkin in a Safeway parking lot. What we do for art!

Yet here I go again, back out on the road, keeping the faith with all those who believe that books still matter, doing my bit to keep the culture alive in this anti-intellectual dark age. I'm looking forward to it: rambling springtime roads in the company of jackrabbits, readers, and quixotic booksellers; rolling through landscapes that my singer/songwriter friends Steve and Kristi Nebel refer to as the "backwoods of bohemia." I've always loved the nickname for the publishing companies' peripatetic reps: book traveler. As a scribbler, not a peddler, "book tourist" suits me just fine.

May/June 2007

NOTE: *For* Mariposa Road, *I undertook the most far-flung and long-lasting book tour I've yet known. However, that might be a thing of the past. Since the sales of e-books exceeded those of trade paperbacks in the spring of 2010, publishing culture has been undergoing even more rapid change. Actual books, and taking them out on the road, may soon be as rare as telephone booths and real letters in the mailbox.*

Condo Picchu

A protected shoreline of mossy balds and maroon madronas stood before me. Sailboats waggled at anchor in the foreground, while white-capped buffleheads bobbed in the bow wake of an incoming ferry. The scene from the dock of the San Juan Island ferry terminal reminded me of a poster I'd spotted days earlier. "There is only so much waterfront," it had said. "Once it's gone, it's gone!" The colorful panorama of unspoiled shoreline had suggested a pitch for a conservation group—maybe the wonderfully successful San Juan Preservation Trust, which recently helped save Turtleback Mountain on nearby Orcas Island. But the poster hadn't been celebrating such things after all. It was just one more ad for one more damn waterfront condominium: *There's only so much, folks. Grab* yours *now!*

I'd been traveling to towns and villages all around the vesiculated shore of Puget Sound, well beyond the bloating extremities of Seattle, and one trait dominated my impression: the extraordinary proliferation of condominium developments, especially near the water. From Port Orchard to Port Townsend, Bellevue to Bellingham, and in many whistle-stops in between, condos are rising like false fronts on a movie set—only these are here to stay, often blocking off the waterfront for everyone else.

Not only up and down the coast, but along the rivers and bays as well, the condos are coming—everywhere, practically, that offers more amenity than a landfill (some indeed are *built over* old landfills). In my own neighborhood, the small city of Astoria, Oregon, known as the West Coast's oldest town, is yet early in this wave—but the tsunami is coming, with five big new projects already permitted and under way along the hitherto accessible shore. On the Washington bank of the Columbia, little burgs from Ilwaco through Skamokawa to our bucolic county seat of Cathlamet are all getting their first acquaintance with condos. Cathlamet's '40s–'50s riverfront served as the backdrop for important maritime scenes in the film *Snow Falling on Cedars* because the filmmakers could not find a suitably period site anywhere else along the coast from Kodiak to Arcata. The condos currently proposed for Cathlamet would erase the town's treasured character overnight. To my astonishment, some people think this would be a *good* thing.

Now, I am not opposed to condominiums in general. Such multi-unit housing can make for efficient use of space rather than consuming endless acres of the hinterlands. And I've seen some marvelous condo conversions that have creatively recycled historic buildings, from silos to schoolhouses, canneries to churches. Architects have brought their best to bear upon some townhouse projects, using green materials and biophilic design to bring green spaces, water, and wildlife into the ken and contact of residents, neighbors, and the by-passing public. I think for example of an array of low- to mid-rise units set back from the Seattle waterfront but close to a sculpture park, beach, and waterside trail, all occupying a former industrial footprint beside the harbor without blocking access to the shoreline itself. Or of certain adobe developments in the Southwest, set modestly into their arid environs, taking a leaf from their Pueblo antecedents rather than ostentatiously upstaging the landscape. No question: with taste and restraint, condos can be done well, respecting their site, their denizens, and the communities of which they will be an undeniable part for decades. I also recognize that condo dwelling simply suits the way many people prefer to live—compactly, near their work, in the city, or in a pleasant and convenient place during retirement. For many, many more people of far lesser means, a rented apartment or high-rise housing is all they can hope to afford beyond the squat, the barrio, or the street.

Condos (and co-ops, lofts, apartments, and other kinds of clustered housing) are a kind of human hive, and the people occupying them are social animals, as are ants, meerkats, and naked mole rats. Whether geometrically precise waxworks like beehives or elaborately interlinked burrows such as anthills and rabbit warrens, social animal colonies achieve relative stability and economy through closely shared housing and intense cooperation. The cooperative abilities of the human animal are not as well developed as are those of social insects or rodents, with many high-rise residents struggling to stay as solitary as possible; and our tendency toward individualism keeps us teetering ever on the edge of chaos despite our best efforts at division of labor. Still, we sometimes achieve temporary social cohesion: communes, collectives, co-housing, and by extension, even cities, can be thought of as evolutionary parallels to the adobe pueblos of mud dauber wasps or cliff swallows. From this standpoint, the proliferation of condos might be seen as an evolutionary adaptation, a move away from space-taking, habitat-fragmenting sprawl, hewing to the elegant examples of more evolved social species. And even though in-filling, that darling of the New Urbanism, co-opts vital vacant lots and open spaces even as it combats sprawl, shouldn't denser dwelling in general trump the old paradigm of spreading out? So bravo for condos, when they really work to the advantage of the people and communities they are intended to serve. But too often the only obvious good accrues to the pocketbooks of the developers, owners, and investors, not to the neighbors, or to the place itself.

Many condos are built as second homes and for pure speculation. They are thrown up to capitalize on the overabundance of cash at the top of the economic heap and the dwindling amount of exploitable open space—especially on the water. After all, there is only so much waterfront…. And because they can be much more profitable than apartments, condominiums replace affordable working- and middle-class housing—an alarming trend in Seattle and many other cities—while raising taxes for everyone around. Nor are these high-end second homes at all like the little mountain or lakeside cabins that Americans have always aspired to own. They monopolize space and impose upon a town's infrastructure while the owners seldom appear, contributing no children to the schools and little to the local economy. Few of the new investor-owners will shop on the main street, until it becomes so gentrified that it serves poorly as a hometown heartbeat. These condo picchus not only sacrifice the nature of particular places, they parasitize the very economies that the city fathers and mothers intended to stimulate through tax exemptions and serving up permits on a silver platter.

It is easy to see why people want to get "theirs" when it comes to waterfront. The late poet Robert Sund, who wrote in a wooden shack on stilts over a back channel of the Skagit River estuary downstream from the ramshackle artists' colony known as Fishtown, expressed water's undeniable draw. "We go to the banks of running streams," he wrote, "though we wash our hands in water every day." But this universal urge can be satisfied by *visiting* a shoreline owned by a park or a land trust, rather than barricading the shore with yet more privately owned Lego blocks. Another poet, Korean writer Ko Un, has said that we will never ameliorate the political and economic strains between cultures until we adopt a posture of "minimum ownership." Nothing could be more opposed to this saving concept than the epidemic mania for *getting*, and getting *more*. And nothing exemplifies such maladaptive behavior better than absentee ownership and speculative, extravagant, wastefully sited lodging that no one really needs.

Few other animals indulge in building "second homes" that sit largely unused. Among the exceptions are several kinds of wrens that construct incomplete, satellite nests beyond the ones where they actually rear their young. According to various ornithologists, these nests function to attract females more effectively, to intimidate competing males, or to confuse predators in search of eggs and nestlings. Perhaps it is not unapt to note that these supernumerary houses have their own name: *dummy nests*.

July/August 2007

License to Kill

Only eighteen wood ticks: not bad, after a long May day's birding in West Virginian woods. The first, adorning my sleeve during beers in the conference center lounge, and the last, plucked from my neck on the way to dinner (after a shower!), went free. But the others, gathered in a tumbler during my tick-check, joined the fetid flow toward the sewage treatment plant. I considered walking them back to the woods, but having already disrobed and de-ticked, I declined. Had they been deer ticks, releasing them would have been reckless in a time of spreading Lyme disease: as one East Coast friend says, "Kill 'em! And kill their filthy spirochetes too!" But even though wood ticks are relatively innocuous, it didn't seem considerate to pop them right outside the window of Rachel Carson Lodge. So, with more misgivings than most people would consider reasonable, I flushed them down the toilet.

My action was optional, but many takings are not. To live, we must kill. Almost all animals do it, even detritivores such as clown fish, cellulose feeders like termites, and dead-matter converters like dung and carrion beetles, all of which wreak collateral damage on microorganisms. Herbivores take life just as carnivores do, only vegetable rather than animal. Pruning can enhance plants, as when light grazing strengthens roots; but heavy foragers, like mountain pine beetles in outbreak mode, may reduce viability, even unto death. Plant predators that consume chiefly leaves, fruits, pollen, nectar, or other expendable parts still gobble a certain by-catch of bacteria and tiny larvae. The fact is, it is difficult to occupy space as a living being without taking the lives of others—altogether impossible for large animals such as we.

Some principled people have tried to opt out of the contract of life living off life. There have been the Albert Schweitzers and the storied Jains, followers of a dharmic faith who believe all living things possess souls and should be held in equal regard. Some Jains reputedly wear masks to avoid breathing microbes and sweep the path before them to prevent the deaths of any insects on which they might tread. We all know people who "wouldn't hurt a flea," at least as far as volition goes. But volition goes only so far. All such conceits are ultimately in vain, if the objective is to kill *nothing*. Thoughtful people can take fewer lives than those who stomp every spider without a thought, or worse. Yet there remains the matter of exerting our weight upon the earth, so richly populated with tiny lives. Of eating, whether cattle, pigs, or krill, grains, beans, or spuds, butchered or harvested by the diners or their paid proxies. Of clothing, since cultivating cotton and hemp means displacement (read: killing) of prior residents with pesticide and plow, and the harvest exerts its own toll. And of shelter, because putting up houses means taking down trees. Our transportation, whether by Hummer or Prius, train or plane, mines living mountainsides for metals. Communications and energy? Try mountaintop removal for coal, open-pit mines for copper, salmon-stream dams, and the entire oil imbroglio, none of them noted for their nonlethal qualities.

But what I want to get to is the killing we do directly—not by dint of mere existence, but of our own volition. We may hunt beasts, rototill worms, dig or spray weeds, slap mosquitoes, trap mice, or swallow—study the word—antibiotics. We each have our own threshold for taking life. A person at one end of the continuum might join the Jains, pay dues to PETA, or espouse (and try to live up to) some Schweitzerian "reverence for life." Organic gardeners cause much less mayhem than conventional growers do with their entomophobic, slay-and-spray ways, but the pesticide-free still dispatch slugs and slaughter chard. Fly-swatters come next, then maybe crawdad-catchers, lobster-eaters, and clam-diggers. Those who can get past the backbone and vertebrate eyes might fish, or go whole hog with warm blood by shooting game birds, squirrels, or deer. From there it's quite a leap to trophy hunters, like our acquaintance who dimmed last year's Christmas spirit with a card depicting himself, his wife, his gun, and the African lion he'd just shot.

So-called "pest control" might start with termites and roaches and move on to mice, moles, packrats, coyotes, and then to shooting wolves from planes, again legal in Alaska after years of prohibition. Ultimately, inexorably, perhaps inevitably, after wolf pluggers and whalers, come the real "thou-shalt-not" killers in the Ten Commandments sense—those who willingly take the lives of members of their own species. These are the warriors, mercenaries, assassins, hit-men, murderers, and executioners—all practitioners of homicide. And finally, at the far other end of the gradient, genocide. Maybe most of us stand closer to Albert Schweitzer than to Albert Speer or Adolph Hitler. But if anyone exists wholly apart from this chain of life and death, then saints still walk.

Humans, being (in their own minds) the ethical animals, have now and then attempted to stem killing, with some success: no official executions in most countries (though plenty in Texas), no more leghold traps in Washington State. If Moses's "thou shalt not" represents the genesis of human efforts against lethality, maybe vegans and anti-capital-punishment activists hold the line today. Surely reducing mortality in the world is a good thing. But just who is throwing stones at the stoners? Animal rightists releasing caged animals such as mink have sometimes caused high losses among the native wild animals in the vicinity. Anti-abortionists have killed doctors with bombs.

One of the more recent anti-lethal lobbies decries the use of insect nets in favor of a strictly observational approach; never mind that I've seen butterfly watchers carelessly trample larval host-plants with caterpillars aboard, or that recent research in Illinois demonstrates that cars kill vastly more butterflies than all the collectors combined. The no-net folks insist that butterfly catching is obsolete, and there's nothing left to be learned that way. This, at a time when one of our most needful goals, the documentation of biodiversity before it becomes extinct, is still a distant dream. And when another of our most baleful problems is nature-deficit disorder, bug nets are one of the cheapest, most effective means of getting kids away from their gizmos and outdoors, into direct contact with nature. Insects reproduce prodigiously, so depleting all but the rarest species with an aerial net is about as likely as eradicating West Nile virus with a fly swatter. Besides, an all-optical approach to insect study is elitist, since few children can afford close-focusing binoculars and cameras. And anyway, kids' attention lasts but a few minutes with lenses, but give them nets and watch them go! Had

E. O. Wilson been forbidden his youthful insect collection, would we even have the concept of biophilia today?

Most of my own butterfly study has long been satisfied by binoculars or catch-and-release, but I continue to collect (that is, dispatch) the odd specimen or two for identification or as distributional vouchers. Coming home from a recent trip to Seattle, I sampled a pair of ochre ringlets for a long-term study of this butterfly and its status. Some of the subspecies of *Coenonympha tullia* proliferate in disturbed habitats, while others, more specialized, are in decline. How we understand them may help to determine their future. Two little ringlets, the color of a Dreamsicle, sacrificed for my idea of science. A bull moose with a big rack might be another man's idea of a good day out, or dinner. I can't say whether one kind of killing is okay, the other not.

We each have a license to kill, by virtue of being born. Humans might be the only animals that worry about how they exercise this elemental act. But in calculating just whom we are willing to kill, and toward what ends, we should ever be mindful of all the lives lost on our behalf. And since we are all complicit, maybe we shouldn't be too quick to condemn the lethal choices of others.

September / October 2007

Pulling the Plug

In the spring of 1969, my Goodwill TV bit the dust. I never got around to replacing it. My household today contains a television set, but it plays only movies. There's no cable or aerial reception here, and we have no dish. I've been without television for nearly forty years, and I've never been sorry. When my stepchildren, Tom and Dory, lived with us, they adapted to the absence of ESPN, cartoons, and *Magnum, P. I.*, and we all read books together before the wood stove each evening. I felt we were living an improbable but lovely idyll.

An adulthood without TV has left me not only culturally spotty but also unselective, so that in a hotel room, if I accidentally turn the thing on, I am in danger of becoming a complete television slut. In the red-eyed morning after, I rue the loss of the sweet night's sleep, the chapter read, the dreams unvisited by drivel and mayhem. Though I fondly remember sitting around our first big set in the '50s, watching Zorro and Cosmo Topper, and my first huge crush on Gail Davis as Annie Oakley, I never regret releasing my home from television's thrall. I'm not smug about it; I'll shamelessly descend on televisioned friends for an election or beg a tape of an HBO concert, and I've become addicted to both *Dallas* and *The West Wing* when living in normally equipped homes for a while. It is recognition of this weakness that keeps me tuned out: I am just not disciplined enough to live with this device. Oh, I can exercise discipline when I really want to—less beer, more exercise, and the ritual of a daily walk, if only to the mailbox or the compost. But the immediate presence of the source of temptation always tests your resolution more keenly.

So, like a bad boy who wouldn't eat his sprouts, no TV for me.

Today there are so many more screens, so many electronic appurtenances, appliances—or say *appendages,* given the way the body politic has been jacked into them. This is not news: the suzerainty of the cell phone, the implant of the iPod, the bramble patch of the Blackberry, the prosthesis of the Palm Pilot, and now the whole enchilada rolled into the iPhone—it's all old hat, as new as it is. In fact, the condition of human bondage to gizmos is itself an old condition. The most common postures today, cell phone elbowed to the ear or digital camera held at arm's length as if intermediary to the actual world, may be recent evolutionary mutations, but that doesn't mean there were no precedents. In Charles Frazier's novel *Thirteen Moons*, protagonist Will Cooper laments the imposition of the telephone in his home: "So urgent, like a watchman sounding a fire alarm, but surely false in the shrill report of the tiny hammer beating frantically against the two acorn-shaped bells. What message short of disaster could be so pressing as to require that horrible jangle?" (If he could only hear the array of irritating ringtones today!) "Use the post," Will advises, "and learn the virtues of patience and silence." This, nearly a hundred years ago!

But some kind of evolution is taking place, and cannot be ignored even by the most resolutely backward among us. At a recent family wedding, my brother-in-law Leon challenged my concern over all the switched-on kids, swapping (as I saw it) the rites of fort building and crawdad catching for the rights of a high-speed wireless connection. "How do you know," Leon asked me, "that these kids aren't just as stimulated, and ultimately fulfilled, as we were by making up our own games outdoors?" I had to admit that he had a point. How indeed could I be sure? But by the third glass of champagne I had an answer—or at least a couple of questions: For one, what happens to a species that loses touch with its habitat? And where will all the conservationists come from when kids no longer have a patch of ground that they can truly call *my space*?

I've been thinking along these lines partly because e-mail has been driving me bats. The only time I can actually feel my normally underachieving blood pressure rise is before the e-screen, inbox at 98 percent full or so, more spilling in by the hour, chiefly importuning, and no chance of ever appeasing it all. And unlike a letter, there's no grace period with e-mail: send one, and the answer comes back in a flash, demanding another *right now.* This creates masses of unnecessary work and entropy, which is what the *e* in e-mail really stands for, but worse is sitting butt-stapled to the swivel chair, eye-sutured to the flat screen, as the pernicious electroglyphs strike the terms of my existence. Years ago I took my home offline because writing, hearth, and health proved incompatible with having e-mail under my roof. Twice a week, I've been going online at the village computer center a few miles from home.

I know I am missing out on some wonderful exchanges and capabilities. But I already weep over all the indoor hours when I could actually be *out,* combing the moss for waterbears or contemplating the profound mystery of where people get the time to read *blogs,* for gods' sakes—is it at the complete expense of books? If I had a mobile phone, I could be available for anyone to reach, anytime ... except, as Greg Brown sings, "You can try me on the cell, but most places I want to be, it don't work."

I suspect that the mass capture of our synapses by electronics may threaten not only serenity but society itself. On a recent train trip, as I was writing with pencil on paper, with one eye out the window on yellow-headed blackbirds and paint foals, I saw something that appalled me: a youngish mother, supplicating babe in one arm, the other grafted to a cell phone on which she was playing a video game. The device went on and on, zinging, pinging, and ringing away, as the baby begged for its mother's presence. She'd pause a stroke to shove a chip into the child's mouth, or tell it to watch the passing lights, but she never looked it in the eyes. "You'll drive everyone crazy if you keep on crying," she scolded. I told her that it was the noises from her machine that were driving all of us crazy. "Oh, this?" she said, and muted it, but kept on playing into the night. I wanted to add, "... *and* your rotten excuse for mothering." Then the scene repeated itself with a different mother, a different baby, in the Sacramento station. These mobile moms brought to mind experiments with baby monkeys given sock dolls for surrogate mothers, and how the babies became sociopathic. This could be even worse: sibling envy for a Nokia.

Back in that lively spring of 1969 when the tube gave out, I certainly had no business sitting sessile before a screen. Now, having just entered what will be, at best, the last quarter or so of my life, I have even less. The last words left for his friends by a fine northeast Oregon naturalist, writer, and man, Frank Conley, felled way too soon by melanoma, drove this understanding home for me: "There will be a memorial service every day you take yourself, or someone else, out into the great outdoors—away from the monitors, videos, and TVs we see mirrored in our eyes—and learn something about birds, butterflies, or biscuitroot." Amen, Frank.

So I'm going to do it—to pull the e-plug, although I'll probably do a Web search now and then at the computer center or the library. Time will tell whether I can make a living without e-mail. In the meantime, I'm going back to the post, and the virtues of patience and silence. My loss, you'll say. Maybe so. We'll see.

November/December 2007

NOTE: It hasn't quite worked out that way. I gave it a good run, going off-line almost altogether throughout 2008. But as I feared, making a freelance living these days without e-mail proved undoable for me. Still, since the analog interregnum, I've greatly changed my e-habits, so that today I am able to keep it down to a dull roar. I write many letters by hand and do my best to help keep our post office alive. We get movies by mail instead of downloading them, and my telephone still has a cord and a dial. When I am not using this machine for my actual work of writing, I try to be elsewhere—in a book, with a person, out of doors—anywhere else at all.

Overseer of Butterflies

During a recent visit with my older brother's family in Colorado, I asked Tom if he was still working at a computer-shop job that he'd held for some years to supplement his photography business. "Nah," he said. "Once I got my government job, I quit there."

"What government job?" I asked. This was news to me.

"Trail inspector," he said.

"What, for the Forest Service?"

"Yep," he said. "I check out the condition of the trails, and the government sends me a paycheck."

I knew Tom had been getting out on the national forests even more than usual lately, making some ambitious hikes in search of downed World War II–era aircraft and taking up mountain biking on top of his longtime devotion to motorized trail bikes. So this seemed plausible, except that I'd never heard of such a job for the USFS, which in any case has been so starved under Bush's budget that staff have been jettisoned like autumn leaves in a gale. Even trail maintenance has devolved largely to volunteers.

"So they pay you to do what you'd like to be doing anyway?" I asked.

"Darn right," Tom confirmed. His wife, Mary's, expression was something between a smirk and *Yeah, right!*, and I figured there must be more to the story.

Tom went on: "The paycheck comes from the Social Security Administration, but the title of the job is Trail Inspector."

I loved the way Tom construed his rightful Social Security benefit, and how he defined the job he undertook to perform with its support. It reminded me of Henry David Thoreau, in *Walden*. When I returned home I found the relevant passage on page sixteen of my annotated 1995 Houghton Mifflin edition, in the chapter called "Economy": "For many years I was self-appointed inspector of snowstorms and rain-storms, and did my duty faithfully; surveyor, if not of highways, then of forest paths and all across-lot routes, keeping them open, and ravines bridged and passable at all seasons, where the public heel had testified to their utility." Of course, Thoreau also worked as an actual surveyor, a pencil maker, and an occasional laborer, presumably all paid positions. But when it came down to describing his dream job, it was "inspector of snowstorms and rain-storms." Thoreau goes on to complain that, after he has faithfully rendered these services for years, the townsmen still decline to "make my place a sinecure with a moderate allowance." How lucky then that Tom, in a similar post, should receive a sanctioned allowance! Come to think of it, our younger brother, Bud, has a similar gig. Disabled by a cattle truck almost forty years ago, he has always worked as a poet, painter, and Neighborhood Observer of his West Denver district. His salary is our late father's Social Security (did I mention that I am a Democrat?).

Don't we all wish for the same: to define our own most desirable employment, and make a living by it? I remember that when George McGovern was the Democratic candidate for president, he called for a guaranteed minimum national income—a pittance, but enough for the likes of me. McGovern imagined that this would relieve the welfare rolls while encouraging all manner of productive activity in the arts and volunteer services. It sounded great: I could have dropped out of the job scene then and there to devote myself to writing and activism. In the end, voters overwhelmingly approved the Republican notion of competing for what you earned—as much of it as you could possibly corner—and that model has clearly prevailed in our culture. For my part, after briefly flirting with regular employment, I pretty much followed McGovern's plan anyway, maintaining the part about the "minimum" but managing always to evade the *guaranteed* bit.

Contemporary society no more encourages such self-definition than did Thoreau's townsmen, unless one's professional aspiration corresponds with what happens to be valued in the marketplace this week, or this year. But just as Tom has found, there is nothing to prevent anyone from designating a primary enthusiasm as an alternate vocation. As long as you can manage to keep body and soul together and muster enough time and energy for it, you can proclaim yourself Manager of Marigolds, even if the Marketing Manager still pays the bills. In fact, I have known many people with day (or night) jobs who have been able to devote more unbroken attention to their passions than those who practice the same activities professionally. When you can leave work at work, your "hobby" time becomes sacred; and then, if you are one of the lucky ones who still has a paid retirement to look forward to, there is no stopping you as that boundless era unfolds. This is one of the reasons that much of the high-level natural history work being done in this country today is the product of amateurs— which means, by the way, "one who loves." Of the three most productive lepidopterists in my state, each of them performing and publishing biology of a high standard, only one is employed as an entomologist; one of the others is retired from an environmental agency, and the third is a working Teamster. But they are all Lepidopterists with a capital *L*.

Others put their energies and their intimate identities to work in the form of volunteerism. The fact that they are spending their "free time" and not being compensated financially takes nothing away from the ultimate payoff such activities provide.

It just so happens that I am about to reclassify my own job title. Oh, I will always be a writer, and an activist. But for the next twelve months, I will not be a speaker, a teacher, a guide, or a consultant, nor will I practice any of the other random trades that have subsidized my primary vocation as scrivener, watcher of slugs, and mumbler through moss. Emulating my big brother's example, as I often did as a boy, I am designating myself Overseer of Butterflies. For the year 2008, I will go forth in Powdermilk, my ancient little Honda (now with 353,000 miles on the odometer), and attempt to encounter and deeply experience as many of the eight hundred species of butterflies that live in the United States and Canada as I can. This will be the first Butterfly Big Year, inspired by the analogous enterprise that birders have undertaken for decades. Kenn Kaufman wrote an unforgettable account of such a quest in his sublime book *Kingbird Highway,* which was inspired by Roger Tory Peterson and James Fisher's *Wild America.* Houghton Mifflin, publisher of both those books, is kindly gambling that the chronicle of my travels will be a worthy successor, though that is a mile-high order.

While I will be tallying the species I see, I fully expect the numbers to take a far back seat to my panoramic view of the land, its condition as habitat, and the way it is changing in our time. Through the compound eyes of the butterflies, I will take a broad look at how these creatures are weathering the changes, and how the warming weather, in particular, is affecting them. It'll be on the cheap (I'm not old enough to collect my Social Security, though you can bet I will as soon as I am) and simple: just my binoculars, my old butterfly net, Marsha, Powdermilk, and me, traveling with a tent, a campstove, and a few bucks for cheap eats and the occasional room in a rundown motor court. The days, the sun, the road, the snowstorms and rainstorms, the butterflies and their plants will be my warp; my weft, the grace and trials of happenstance.

To define the project further would defeat its purpose, which brings to mind a long-ago Washington State election when a self-styled fellow named Richard AC-DC Green ran for Commissioner of Public Lands on the Owl Party ticket. When asked about his platform, Green simply reiterated his campaign slogan: "If elected, I plan to go forth fearlessly and commission the land."

I always thought it a little sad that Richard AC-DC Green was not elected to office. But in my own way, I plan to fulfill his campaign promise for him.

January/February 2008

NOTE: *My year as Overseer of Butterflies is reported in the book* Mariposa Road *(Houghton Mifflin Harcourt, 2010). Since then, I have developed a simple solution for long-term financing of Social Security and Medicare: nonbelievers in Roosevelt's New Deal and Johnson's Great Society need not apply. And now that I am approaching Social Security age myself, I am thinking of resurrecting Roosevelt's Federal Writers' Project. That marvelous WPA program from the Great Depression is long gone, but there's nothing to keep me from making my own New Deal with that future federal paycheck.*

Magpie Song

After the forever-flight from Portland to Perth via San Francisco and Sydney, I slept the sleep of the crypt. It would be weeks before my Circadian rut and I settled in comfortably again together, but there is something about sleep deprivation that heightens the senses, which is why it has been an important element of vision quests. And this boldly patterned entity staring back at me from the lawn outside my window at St. Catherine's College was certainly a vision.

Clearly, this presence was crowlike, shockingly pied in black and white. Its chalky, front-heavy bill reminded me of an English rook as it yo-yoed to worm a niblet. My field guide showed it to be the Australian magpie, belonging to the bell-magpies, but it might as well have been called a crow; after all, when Audubon named the nutcracker of the western peaks for its finder, he called it Clark's crow. But there is already an Australian crow, and an Australian raven, small but making up for it with a bushy beard. So it is not surprising that this Antipodean bird was tagged "magpie," if only for its two-tone suit—black face, back, and breast; white mantle, wings, and belly. What *was* surprising was its beautiful song. Unlike the Australian crow's harsh caw or the raven's ubiquitous squall that sounds like a seriously pissed-off pussycat, the magpie warbles like a whisper, or a flute.

I was in Perth to attend a conference called Come Outside and Play, so when all the talk was over, I did. Along the bush trails of enormous Kings Park, spring wildflowers were peaking. The most prominent pink flowers were the introduced wild gladiolas, but many native plants also contributed to the extraordinary palette. Strangest were kangaroo paws, the state flower—velvety, two-foot-tall, red and green wonders that would be hard to dream up from scratch. This would be the perfect Christmas plant, if only it bloomed Down Under in midsummer instead of spring.

The birds more than matched the fabulous flowers. I never tracked down one of the brilliant blue fairy wrens that all birders covet when they come here, though I did see emus from a bus, and kookaburras with their exaggerated bills, sitting indeed in gum trees. And fantails, and honeyeaters, and unbelievably metallic bronzewings forged from some ornithological alchemy, all thrilling and completely outside my experience. What most rang this birder's bells, however, were the psittacines: the parrots. On my first walk out, ungodly squawks and cascading petals announced the first of many flocks of rainbow lorikeets, stunning wonders like painted buntings stuffed into parrot suits. When you see the purple, blue, scarlet, orange, and wild, dazzling, parrot green as they daub themselves all over the fruit trees, tossing husks here and there with abandon, you know you are not in Kansas anymore—or southwest Washington, for that matter. But the lorikeets are not easy on the ear. They screech in flight like a pack of tractors grinding through their gears in falsetto, making you cover your ears even as your eyes pop open as wide as they'll go.

I was coming around a curve in Kings Park when I heard a pleasant, purring bell-tone that rose into a bright "dotty dot" or several-noted ring. It took me a second to decode, but then I discerned the words "twenty-eight." Of course! This was the 28 parrot, also known as the Australian ring-neck. A dozen or more 28s foraged on the trail before me: grapy head, yellow neck-ring, deep blue wings, and pure chartreuse overall. They remained until I was almost upon them, behaving in a way that the old bird books call "confiding." All through my visit the noisy lorikeets were unavoidable, while the 28s appeared softly, here and there and now and then.

But the parrot that really grabbed my heart was the galah. Plump with rounded white topknots and bills, galahs look baby-faced to me. They also recall a popular color scheme of the 1950s: pink and gray. Their backs and wings are the softest of dove grays, and their breasts are the pink of rhubarb pie, of coral, of the ventral forewing of a Virginia lady butterfly—a hue uncommon in the animal world, more often seen in the realms of petunias or popsicles.

I found a pair of galahs guarding a tree hole in a dead snag on the beautiful University of Western Australia campus. I reckoned that the tree had been left standing for this broody pair. But when I looked a little later, the pair of parrots at the tree hole were 28s; and on a third pass, I found rainbow lorikeets in possession! Do they just trade off in shifts, I wondered, or drive each other away in turn? And which species will end up nesting there? The galahs were hanging about nearby each time the others held the fort, so I was rooting for them, especially after I found the favor of a rich pink plume, a breast feather fallen to earth after an energetic preening or an encounter among rivals. Like a drunken cheerleader at a tailgate party, I chanted to myself, "Go, galahs!" It occurred to me that I could happily come here and engage in a thesis study of their behavior, giving me an academic excuse to simply watch them all day long.

Still, as much as I enjoyed the fresh sensations of Perth, I was distressed by something else I saw, or didn't see: no one else even noticed the parade of the parrots! As I stood beneath the snag watching the changing of the psittacine guard, hundreds of students and faculty and others walked past. Plenty of passersby looked at this obvious outlander with binoculars; I was the exotic element here. But not a single person shifted his or her gaze a few feet to behold this splendid, made-in-Australia avian spectacle. Is it that easy to become jaded to the nature about us, even such a striking expression as this? I guess that's what the conference title, Come Outside and Play, was getting at. The passersby were out of doors, all right, but they certainly weren't playing. Rushing about with cell phones and laptops, they were having no fun at all, and it broke my heart to see them so detached from the world.

Now it's another week, another conference, and I am back in the Northern Hemisphere. I'm not only jet-lagged beyond retrieval, but also without a reference point in time or space. Until, that is, I hear a familiar sound coming from the lawn outside my window at the Aspen Institute—the querulous *maaaag?* of a black-and-white magpie, no less sweet to this Colorado-bred lad than the pretty song of its Down Under namesake. And what's that over there, in the cottonwoods, among blue sage? A flock of redpolls, my favorites! Northern finches come south on winter's breath, redpolls are cherry-crowned and rosy-washed over their snowy breasts with the very pink of this galah's breast feather, tucked away here in my notebook.

These birds situate me, pulling me out of that ungrounded state so familiar to those of us who follow this weirdly vagile way of life called the lecture circuit. Even more important, they remind me of something I hope I will never forget: no matter where I go, if only I'll *look,* I can never be anywhere other than at home in the world. For ten years of Tangled Banks, I've tried to say this in as many ways as I could imagine, and I'll say it one more time, for it bears repeating, like a mantra: just look around yourself, *really look,* and the actual world will never let you down.

So this is the last "Tangled Bank." But it's not my swan song; more, I would say, my magpie song.

March/April 2008

Epilogue: X the Unknown

As kids in the fifties and early sixties, my big brother, Tom, and I loved monster movies. Nothing like the slashers of today, which I can't stand, these were the early Frankensteins and Draculas and Wolfmen and Mummies, starring Boris Karloff, Bela Lugosi, and Lon Chaney, Jr. The violence was only suggested, the personalities of the tragic figures were developed (or so it seemed to our easy sensibilities), and the tension could be exquisite. Certain space movies qualified, like *The Forbidden Planet* with its all but invisible, cat-like monster projected from Walter Pidgeon's id, and a few psycho-thrillers such as *Macabre* and *Split Second* grabbed us too. Of course, in those days of "duck and cover," the ever-present Bomb, and the Red Scare, all manner of radioactive threats crept onto the matinee screen: giant tarantulas, mutated sea creatures, and—most memorable by far to this day—*X the Unknown*. Experiments on a Scottish bog draw forth a radiation-seeking *something* (true to the title, we never find out quite *what*) that melts people's faces, grows, and spreads, as Dean Jagger leads the battle against it.

This movie was a little more graphic, or gross, than most we watched, and very, very scary. But quite aside from the slender plot and chilling black-and-white effects, I think it is the title itself that sticks in my memory. There is something about the idea of the sheer unknown that thrills me, draws me out of doors or into a book, makes me excited to get up and face the day so as to press into territory never before plumbed ... which might be a clump of moss just as well as a new trail or state or mountain range. It's as if things can't be too bad, as long as there are brand new discoveries out there to be made. Do I flatter myself to suppose that Darwin himself was drawn or driven by the same impulse when he took on the pollination of orchids, the formation of coral atolls, the sex of barnacles, the power of movement in plants, the making of soil duff by earthworms, and much more, then wrote great books about each of his preoccupations in turn?

For what better way to come to grips with X, than to write about it? And what, after all, defines the "X" of the matter; what is the actual nature of the "unknown?" It is one thing to approach true terra incognita, somewhere or something that no one has ever before plumbed—or at least published. Few of us will have such opportunities, although they are not all gone: when I found the first monarchs west of the Rockies migrating into Mexico, or the first arctic skippers recorded in Columbia County, these were new quanta added to the sum of our knowledge. But just as righteously to be called unknown is any question we have not answered for ourselves alone. My topics have borne little of the "grandeur" of which Darwin spoke when he referred to "this view of life." I have engaged simple enough ideas, questions, precepts, mysteries, and biases. But with each excursion, I have essayed, in Montaigne's sense of the term, some element of the earth with which I dearly wished to acquaint myself, and discover what I thought about it.

Now that I am no longer writing these short pieces for *Orion*, I miss the exercise, the enterprise, the outright *surprise* of the thing. So I take up my journal, or write a letter, a poem, or even, the gods help me, a contribution to someone's blog. There is no denying that "The Tangled Bank" was a privileged pulpit from which to utter these moss-murmurs and fern-words, more than half a hundred times, and make them fast on paper. It gave me a rare invitation to ask my questions, explore my opinions, and chase my whims to a degree any essayist would, or should, envy.

This lucky gig also offered the rare gift of working intimately with a maddeningly good, sweet-natured, rough-and-tumble editor who almost never let me get away with writing crap. (When her tender years rendered my cultural references occasionally opaque, I usually deferred in favor of youth. But Broderick Crawford was non-negotiable, so I referred Jennifer to Google.) My long engagement with the best magazine going (*no ads at all!*), and the extended conversation with its audience of excellent and caring readers, can only be called an honor.

In the end, "The Tangled Bank" died a natural death, as most columns probably should, before they become decadent. But since I never wrote about anything that failed to fascinate or fuddle me at the time, I sometimes wonder: were I to write another decade's-worth of "Banks," just what briar-patches would I be tempted to leap into, root around in, and untangle to my own satisfaction? Here are fifty-two more unknown X's that spring to mind:

Waterbears, and could I really find some? flies, in general; the history of saffron and the ontology of crocuses; the tickle in the ball of my left thumb at just the right titer of alcohol; communal roosting involving different species, such as elk and geese; plants that grow on animals, such as lichens on sloths and moss on beetles; animals that live only on plants that grow only on animals; the indigenous fauna of saunas, including Finns; Bessemer & blacksmiths, swords & ploughshares, guns & butter: the molecular structure of war; the hummingbirdness of things; how Johnny Winter can be so skinny and play the guitar like that; the evolution of empathy; what really happened at Agincourt, and the various powers of yew trees; the Crusades, the Corps of Discovery, the astronauts—what about jock itch? the natural history of heaven and hell; does high culture depend upon high crimes? Prohibition as extinction event; the sutra of sufficiency, the rhetoric of riches, and how much is enough; the evolutionary opportunities after war; gleaners and scavengers in an age of plenty; how do crab spiders find their perfect perches? the relativity of celebrity; Paganini, Robert Johnson, and the devil; recycling roadkill; Utopia: what to leave in, what to leave out; is money necessary? are pledge drives necessary? what were "sharp-shod animals," anyway? giant flying squirrels, Dreamliners, and the limits of gliding; bookmarks: how readers leave their spoor; the heralded death of correspondence; glimpsing the platypus, or what counts as encounter? succubi and incubi; violence in play and in practice; dreams as movies; movies as dreams; Deuteronomy, dust, and Darwin: the notion of "dirt"; what if the poets ruled? public lands and private lives; why so many willows in music? Valentine, Nicholas, and Christopher: saints with a slant; mink and muskrat, owl and thrush: predation as a double death; The Fall as historical trampoline; Thomas Tallis, Pinetop Perkins, and Bruce Springsteen: that which lasts; do springtails in the soil really outweigh the Sequoias overhead? the Owls of Yale: is Athena home? Diogenes' lantern dims in the glare of the Pearl; The Market vs. mercados; *probability and Pooh-sticks; whither the Kalakala, and why do I care?*

Why do I care about any of this, anyway? As Graham Swift wrote in his novel *Ever After,* "Why is anything special? Either everything is special, which is absurd. Or nothing is special. Which is meaningless." Take your pick!

ACKNOWLEDGEMENTS

Great and heartfelt thanks belong to my friend and neighbor David White, for laboriously creating the digital files from the original printed copy, both for this book and my earlier OSU Press title, *The Thunder Tree*. In generously doing so, he made the whole thing possible.

Jennifer Sahn and Aina Niemela are two of the kindest, toughest, and most astute editors I have ever known. I owe "The Tangled Bank," and thus this book, to them, and to Marion and Olivia Gilliam, George Russell, and all the other *Orion* pioneers and workers who gave me the chance not once, but twice, to pen these hackabouts and whims.

I thank *Orion* editor-in-chief Emerson (Chip) Blake, the directors of the Orion Society, and all my friends at *Orion Magazine* for their enthusiastic support for the republication of these essays as a book. And I thank Laurie Lane-Zucker for the Forgotten Language Tours and Tajikistan, and much more.

The members of my ten-year writing group read and most helpfully critiqued many of these pieces, and I thank them with fervor and affection: Greg Darms, Brian Harrison, John Indermark, Diane Matthews, Susan Pakenen Holway, Bryan Pentilla, Patricia Staton Thomas, Jenelle Varila, and Lorne Wirkkala. Fayette Krause, Peter Dunwiddie, and members of my Kittredge writing class at EVST, University of Montana, all read and commented on individual essays, and I thank them warmly along with any other readers whom I have not named. And our good old mentor Arthur Kruckeberg (ARK) has once again been faithful to the mission.

At Oregon State University Press, again I am deeply obliged to Mary Elizabeth Braun, Tom Booth, Jo Alexander, Micki Reaman, and their colleagues for seeing the light in these essays, and for allowing them to shine in one of the beautiful books they continue to faithfully make in a darkening age for the Book. I thank them hugely, and also Barbara Stafford for her stunning cover art and David Drummond for his cover design.

As always, but never rotely, I thank with all my heart my dear wife, Thea Linnaea Pyle, for her Queen Anne's lace motif on the title page, and all else. She is always my vital editor of first recourse, and the inspiration for everything I write. "The Chemistry Between Us" lives on.